Essential Mathematics for Life

BOOK 4

Graphs, Measurements and Statistics

Fourth Edition

McGraw-Hill

New York, New York
Columbus, Ohio
Mission Hills, California
Peoria, Illinois

Authors

Mary S. Charuhas
Associate Dean
College of Lake County
Grayslake, Illinois

Dorothy McMurtry
District Director of ABE,GED, ESL
City Colleges of Chicago
Chicago, Illinois

The Mathematics Faculty
American Preparatory Institute
Killeen, Texas

Contributing Writers

Kathryn S. Harr
Mathematics Instructor
Pickerington, Ohio

Priscilla Ware
Educational Consultant and Instructor
Columbus, Ohio

Dr. Pearl Chase
Professional Consultants of Dallas
Cedar Hill, Texas

Contributing Editors and Reviewers

Barbara Warner
Monroe Community College
Rochester, New York

Michelle Heatherly
Coastal Carolina Community College
Jacksonville, North Carolina

Anita Armfield
York Technical College
Rock Hill, South Carolina

Judy D. Cole
Lafayette Regional Technical Institute
Lafayette, Louisiana

Mary Fincher
New Orleans Job Corps
New Orleans, Louisiana

Cheryl Gunderson
Rusk Community Learning Center
Ladysmith, Wisconsin

Cynthia A. Love
Columbus City Schools
Columbus, Ohio

Joyce Claar
South Westchester BOCES
Valhalla, New York

John Grabowski
St. Joseph Hill Academy
Staten Island, New York

Virginia Victor
Maple Run Youth Center
Cumberland, Maryland

Sandi Braga
College of South Idaho
Twin Falls, Idaho

Maggie Cunningham
Adult Education
Schertz, Texas

Sylvia Gilliard
Naval Consolidated Brig
Charleston, South Carolina

Eva Eaton-Smith
Cecil Community College
Elkton, Maryland

Fabienne West
John C. Calhoun State Community College
Decatur, Alabama

Send all inquiries to:
Glencoe/McGraw-Hill
936 Eastwind Drive
Westerville, Ohio 43081

ISBN: 0-02-802611-X

2 3 4 5 6 7 8 9 POH 02 01 00 99 98 97 96

CONTENTS

Graphs, Measurements, & Statistics

Unit 1 Review of Decimals, Fractions, Percents, and Proportions

Unit 2 Measurement

Unit 3 Picture Graphs and Tables

Unit 4 Bar Graphs and Line Graphs

Unit 5 Circle Graphs

Unit 6 Statistics

Unit 7 Probability

UNIT 1

Review of Decimals, Fractions, Percents, and Proportions

Pretest

Solve the following.

1. $\begin{array}{r} 10.831 \\ +\ 4.09 \\ \hline \end{array}$

2. $\begin{array}{r} 126.72 \\ -\ 33.164 \\ \hline \end{array}$

3. $\begin{array}{r} 78.02 \\ \times\ 2.6 \\ \hline \end{array}$

4. $0.3\overline{)644.7}$

5. $\begin{array}{r} 3\frac{2}{5} \\ +\ 5\frac{7}{10} \\ \hline \end{array}$

6. $\begin{array}{r} 17\frac{2}{3} \\ -\ 2\frac{3}{4} \\ \hline \end{array}$

7. $4\frac{2}{5} \times \frac{5}{14}$ __________

8. $4\frac{1}{2} \div 2\frac{4}{7}$ __________

Write the equivalent of each of the following.

	Fraction	Decimal	Percent
9.	$\frac{1}{5}$	____	____
10.	____	0.06	____
11.	____	____	54%

	Fraction	Decimal	Percent
12.	______	1.3	______
13.	$\frac{3}{8}$	______	______
14.	______	______	0.2%

Compare the following, using <, >, or =.

15. 0.02 ______ 20%

16. $4\frac{7}{8}$ ______ 4.75

17. 16% ______ $\frac{1}{4}$

18. $\frac{3}{5}$ ______ 0.6

19. $\frac{9}{100}$ ______ 90%

20. 5% ______ 0.005

Problem Solving

Solve the following problems.

21. If a sale on bananas is 3 pounds for $0.99, how much will 5 pounds cost? ______________

22. What is the interest on a $45,000 home loan at 8% interest for 25 years? ______________

23. One percent of the entire staff at the company retired this year. If there are 9 retirees, how many people work for the company? ______________

24. Last week, Chelsea bought a jacket that cost $36. This week she saw the same jacket on sale for $27. What was the percent of decrease? ______________

25. In the hotel 40% of the rooms were reserved for nonsmoking patrons. If there were 315 rooms, how many were for nonsmokers? ______________

LESSON 1

Adding and Subtracting Decimals

When adding or subtracting decimals, follow these steps:

Examples

A. Add.
3 + 4.05 = ?

B. Subtract.
15.07 − .005 = ?

Step	Instruction	A	B
Step 1	Write the numbers in a column, lining up the decimal points. Place a decimal point to the right of whole numbers.	3. + 4.05	15.07 − .005
Step 2	Add zeros to the right of the decimal point, if needed.	3.00 + 4.05	15.070 − .005
Step 3	Add or subtract as with whole numbers.	3.00 + 4.05 7.05	15.070 − .005 15.065
Step 4	Bring the decimal point straight down into the answer.		

MATH HINT

To check addition, add in the other direction, or subtract. To check subtraction, add.

Practice

Add or subtract.

1. 211.6 + 4.19 ________

2. 8 − 2.37 ________

3. 1.97 − 0.98 ________

4. 476.1 + 2.5 + 40.08 ________

5.
$$\begin{array}{r} 13.65 \\ -\ 0.17 \\ \hline \end{array}$$

6.
$$\begin{array}{r} 451 \\ 6.89 \\ +\ 0.35 \\ \hline \end{array}$$

7.
$$\begin{array}{r} 17.04 \\ -\ 3.261 \\ \hline \end{array}$$

Problem Solving

Solve the following problems.

8. The total finishing time for three runners to finish a marathon was 11.24 hours. If two of the runners' times were 3.52 and 3.88 hours, what was the finishing time for the third runner? ________

9. On Tuesday, it snowed 4.35 inches; on Wednesday, 9.15 inches; and on Thursday, 7.8 inches. If the total snowfall for Tuesday through Friday was 22.8 inches, how much did it snow on Friday? ________

10. Paula bought a skirt for $29, a blouse for $22.50, and a sweater for $35.95. If the tax was $4.37, what was the change from a $100 bill? ________

Problems 11 and 12 are related.

11. Last week, Vince worked 9.5 hours of overtime. This week, he worked 7.9 hours of overtime. How many more hours of overtime did Vince work last week than this week? ________

12. How much overtime will Vince need to work next week to have 20 hours of overtime in all? ________

Multiplying Decimals

When multiplying decimals, follow these steps:

Step 1 Multiply as with whole numbers.

Step 2 Count the number of decimal places in both numbers and add them together.

Step 3 Count off this total number of decimal places in the answer. Count from right to left.

> **MATH HINT**
>
> If the answer does not have enough digits to place the decimal point in the correct place, add zeros to the left as placeholders.

Examples

A. Multiply.
$50 \times 3.2 = ?$

Step 1

$$\begin{array}{r} 50 \\ \times\ 3.2 \\ \hline 100 \\ 150 \\ \hline 1600 \end{array}$$

Step 2

$$\begin{array}{r} 50 \\ \times\ 3.2 \\ \hline \end{array} \qquad \begin{array}{r} 0 \text{ places} \\ +\ 1 \text{ place} \\ \hline 1 \text{ place} \end{array}$$

Step 3

$$\begin{array}{r} 50 \\ \times\ 3.2 \\ \hline 160.0 \end{array} \qquad 1 \text{ place}$$

The answer is 160. (The zero in the tenths place is not needed. It may be dropped.)

B. Multiply.
$1.025 \times .006 = ?$

Step 1

$$\begin{array}{r} 1.025 \\ \times\ .006 \\ \hline 6150 \end{array}$$

Step 2

$$\begin{array}{r} 1.025 \\ \times\ .006 \\ \hline \end{array} \qquad \begin{array}{r} 3 \text{ places} \\ +\ 3 \text{ places} \\ \hline 6 \text{ places} \end{array}$$

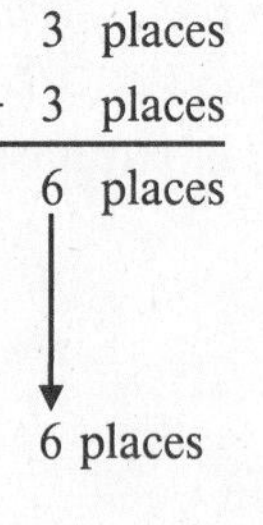

Step 3

$$\begin{array}{r} 1.025 \\ \times\ .006 \\ \hline .006150 \end{array} \qquad 6 \text{ places}$$

The answer is .00615. (The zero in the millionths place is not needed. It may be dropped.)

Practice

Multiply.

1. $\begin{array}{r} 369 \\ \times \ .07 \\ \hline \end{array}$

2. $\begin{array}{r} 21 \\ \times \ 6.5 \\ \hline \end{array}$

3. $\begin{array}{r} 5{,}492 \\ \times \ .101 \\ \hline \end{array}$

4. $\begin{array}{r} .035 \\ \times \ 41 \\ \hline \end{array}$

5. $\begin{array}{r} 12.98 \\ \times \ .02 \\ \hline \end{array}$

6. $\begin{array}{r} 78.13 \\ \times \ 2.7 \\ \hline \end{array}$

7. 664 × 1.95 ________________

8. 1,000 × 0.43 ________________

Problem Solving

Solve the following problems.

9. The price of a movie ticket is $7.50 for an adult and $4.25 for a child. What is the cost for a family of three adults and two children? ________________

10. Jack's Sound Shack sells compact CD disks for $9.99 each. What is the cost of five CDs? ________________

11. Alejandro charges $44.50 an hour to repair computers in his shop. What would he charge for a job that takes 0.5 hours? ________________

LESSON 3

Dividing Decimals

When dividing decimals by **whole numbers**, follow these steps:

> **MATH HINT**
>
> Remember the parts of a division problem:
>
> 8 (quotient)
> (divisor) 4)32 (dividend)

Examples

Step 1 Place a decimal point in the quotient directly above the decimal point in the dividend.

Step 2 Divide as you would with whole numbers.

A. Divide.

4)8.4

Step 1: 4)8.4 (decimal point placed in the quotient above the decimal point)

Step 2:
2.1
4)8.4
8
4
4

B. Divide.

7)1.75

Step 1: 7)1.75 (decimal point placed in the quotient above the decimal point)

Step 2:
.25
7)1.75
1 4
35
35

When dividing decimals by **decimals**, follow these steps:

Step 1 Make the divisor a whole number by moving the decimal point to the right of the last digit.

Step 2 Move the decimal point in the dividend to the right the same number of places. Add zeros if necessary.

Step 3 Place a decimal point directly above in the quotient.

Step 4 Divide as with whole numbers.

C. Divide.

.05)2.5

Step 1: .05.)2.5

Step 2: 5)2.50.

Step 3: 5)250.

Step 4:
50.
5)250.
25
0
0

D. Divide.

.08).0016

Step 1: .08.).0016

Step 2: 8).00.16

Step 3: 8)00.16

Step 4:
.02
8)00.16
0
16
16

Practice

Divide.

1. $186 \div 0.3$ __________ **2.** $52 \div 0.25$ __________ **3.** $37.8 \div 0.9$ __________

4. $0.08\overline{)720.96}$ **5.** $0.006\overline{)10.2}$ **6.** $6\overline{)0.057}$

Problem Solving

Solve the following problems.

7. The walkway around the lake is 1.32 miles. If the park district places 12 benches at equal distances around the lake, how far apart will the benches be?

____

8. Rachel had a sound system wired into her new home. The job required 1,374 feet of speaker wire. If she paid $233.58 for the wire, what was the cost per foot?

____

9. Theresa prepared 4.75 pounds of taco meat for a party. She expected that each taco salad would require 0.25 pounds of meat. How many salads could she serve?

____

LESSON 4

Fractions

A fraction is part of a whole unit.

$\frac{1}{2}$	numerator	indicates how many parts are being used
	denominator	indicates the number of parts that make the whole

Types of Fractions

Proper Fractions	Improper Fractions	Mixed Numbers
The numerator is smaller than the denominator.	The numerator is larger than or equal to the denominator.	The sum of at least one whole unit and a fraction.
The fraction is less than one whole unit.	The fraction is greater than or equal to one whole unit.	
$\frac{1}{2}$ $\frac{2}{5}$ $\frac{6}{53}$	$\frac{3}{2}$ $\frac{5}{5}$ $\frac{8}{7}$	$3\frac{1}{4}$ $1\frac{9}{10}$ $88\frac{2}{5}$

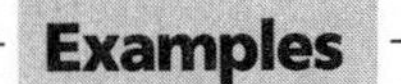

Examples

A. Reducing fractions
To reduce a fraction to its lowest terms, divide the numerator and the denominator by the same number.

$\frac{4 \div 4}{12 \div 4} = \frac{1}{3}$ $\frac{5 \div 5}{15 \div 5} = \frac{1}{3}$

B. Equivalent fractions
To make an equivalent fraction, multiply the numerator and the denominator by the same number.

$\frac{1 \times 7}{3 \times 7} = \frac{7}{21}$ $\frac{2 \times 4}{5 \times 4} = \frac{8}{20}$

Practice

Reduce each fraction or mixed number.

1. $\frac{18}{81}$ ______ **2.** $\frac{5}{70}$ ______

3. $12\frac{14}{49}$ ______ **4.** $5\frac{18}{30}$ ______

5. $18\frac{6}{18}$ ______ **6.** $24\frac{24}{120}$ ______

Write equivalent fractions.

7. $\frac{2}{3} = \frac{\quad}{15}$ **8.** $\frac{4}{7} = \frac{\quad}{28}$

9. $\frac{3}{8} = \frac{\quad}{56}$ **10.** $\frac{1}{4} = \frac{\quad}{92}$

Example

C. Finding the least common denominator
Find the smallest number that is a multiple of all the denominators.

$\frac{1}{3}, \frac{1}{2}, \frac{1}{4}$ 12 is the smallest multiple of 3, 2, and 4.

12 is the least common denominator.

MATH HINT

A multiple is the product of a number and any whole number.
Multiples of 3:
0, 3, 6, 9, 12, 15, . . .

Practice

Find the least common denominator for each set of fractions.

11. $\frac{3}{4}, \frac{2}{7}$ ___

12. $\frac{1}{6}, \frac{1}{3}$ ___

13. $\frac{4}{5}, \frac{8}{11}$ ___

14. $\frac{1}{2}, \frac{3}{4}, \frac{5}{6}$ ___

15. $\frac{1}{3}, \frac{1}{5}$ ___

16. $\frac{2}{9}, \frac{1}{2}$ ___

17. $\frac{1}{2}, \frac{5}{6}, \frac{4}{9}$ ___

18. $\frac{1}{6}, \frac{1}{7}, \frac{1}{8}$ ___

Problem Solving

Solve the following problems.

19. This morning, Anthony delivered the newspaper to 48 of his 52 customers. What fractional part of his customers did not receive the newspaper? Reduce the fraction to lowest terms. ________

20. Among the fractions $\frac{3}{5}$, $\frac{3}{4}$, and $\frac{2}{3}$, which is equal to $\frac{36}{54}$? ________

21. A recipe for breakfast bars calls for $\frac{3}{4}$ cup of rolled oats, $\frac{2}{3}$ cup of granola, and $\frac{1}{8}$ cup of molasses. Write these fractions using the least common denominator. ________

LESSON 5

Multiplying Fractions

To multiply fractions, follow these steps:

Step 1 Change all mixed numbers to improper fractions.

Step 2 Cancel if possible.

Step 3 Multiply straight across.

Step 4 Reduce the answer to lowest terms.

Examples

A. $\frac{4}{5} \times \frac{2}{3} = ?$

$\frac{4}{5} \times \frac{2}{3} = \frac{8}{15}$ Multiply straight across.

B. $\frac{4}{9} \times \frac{9}{16} = ?$

$\frac{\overset{1}{\cancel{4}}}{\underset{1}{\cancel{9}}} \times \frac{\overset{1}{\cancel{9}}}{\underset{4}{\cancel{16}}} = \frac{1}{4}$ Cancel. Multiply straight across.

C. $2\frac{3}{8} \times 1\frac{5}{7} = ?$

$\frac{19}{8} \times \frac{12}{7} =$ Change mixed numbers to improper fractions.

$\frac{19}{\underset{2}{\cancel{8}}} \times \frac{\overset{3}{\cancel{12}}}{7} =$ Cancel.

$\frac{57}{14} = 4\frac{1}{14}$ Multiply straight across. Reduce.

Practice

Multiply. Reduce answers to lowest terms.

1. $\frac{1}{4} \times \frac{1}{5}$ ________

2. $\frac{2}{3} \times \frac{5}{6}$ ________

3. $\frac{1}{3} \times \frac{5}{7} \times \frac{9}{10}$ ________

4. $36 \times \frac{7}{8}$ ________

5. $4\frac{2}{3} \times \frac{2}{3} \times \frac{9}{14}$ ________

6. $\frac{9}{21} \times \frac{7}{27}$ ________

7. $4\frac{1}{6} \times 1\frac{1}{5}$ ________

8. $\frac{3}{4} \times 10 \times \frac{8}{27}$ ________

9. $2\frac{6}{7} \times \frac{3}{4}$ ________

10. $7\frac{4}{5} \times \frac{2}{3} \times 6\frac{1}{4}$ ________

Solve the following problems.

11. The costume committee had $10\frac{1}{2}$ yards of material. Two-thirds of the fabric would be used for dresses and the rest for decoration. How many yards would be used for the dresses?

______

12. The trim for one shelf is $3\frac{1}{8}$ feet long. If there are 4 shelves, how many feet of trim will be needed?

______

13. Three-fourths of the 72 people who took the GED math test received a score above 45. How many people scored above 45?

______

LESSON 6

Dividing by Fractions

To divide by fractions, follow these steps:

Step 1 Change all mixed numbers to improper fractions.

Step 2 Invert the divisor. Change the division sign to a multiplication sign.

Step 3 Cancel if possible.

Step 4 Multiply straight across.

Step 5 Reduce the answer to lowest terms.

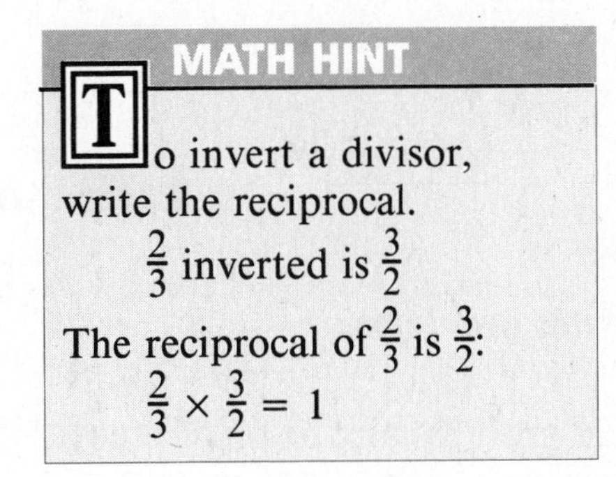
MATH HINT

To invert a divisor, write the reciprocal.

$\frac{2}{3}$ inverted is $\frac{3}{2}$

The reciprocal of $\frac{2}{3}$ is $\frac{3}{2}$:

$\frac{2}{3} \times \frac{3}{2} = 1$

Examples

A. $2\frac{1}{3} \div \frac{7}{8}$

$\frac{7}{3} \div \frac{7}{8}$ Change the mixed number to an improper fraction.

$\frac{\overset{1}{\cancel{7}}}{3} \times \frac{8}{\underset{1}{\cancel{7}}}$ Invert the divisor. Change the sign to multiplication. Cancel.

$\frac{8}{3} = 2\frac{2}{3}$ Multiply straight across. Reduce.

B. $4\frac{2}{5} \div 2$

$\frac{22}{5} \div \frac{2}{1}$ Change the mixed number to an improper fraction.

$\frac{\overset{11}{\cancel{22}}}{5} \times \frac{1}{\underset{1}{\cancel{2}}}$ Invert the divisor. Change the sign to multiplication. Cancel.

$\frac{11}{5} = 2\frac{1}{5}$ Multiply straight across. Reduce.

Practice

Divide. Reduce answers to lowest terms.

1. $\frac{3}{10} \div \frac{1}{2}$ __________

2. $\frac{11}{14} \div \frac{11}{14}$ __________

3. $9\frac{1}{6} \div \frac{2}{3}$ __________

4. $\frac{3}{4} \div 1\frac{1}{2}$ __________

5. $24 \div \frac{10}{13}$ __________

6. $4\frac{5}{8} \div 37$ __________

7. $8\frac{3}{7} \div \frac{1}{7}$ __________

8. $\frac{5}{6} \div 1\frac{2}{5}$ __________

9. $2\frac{1}{3} \div \frac{7}{8}$ __________

10. $6\frac{1}{4} \div \frac{5}{8}$ __________

11. $\frac{1}{9} \div 3$ __________

12. $8\frac{2}{3} \div 1\frac{6}{7}$ __________

Problem Solving

Solve the following problems.

13. A local civic group plans to plant trees along $\frac{3}{4}$ of a mile of highway. This section begins at a stop sign and ends past the city limits. If the group plants 8 trees and does not plant one at the stop sign, how far apart will each tree be?

____

14. A concession stand has 500 gallons of lemonade to sell at the baseball game. If each large dispenser holds $5\frac{1}{3}$ gallons, how many dispensers could be filled?

____

15. Florence sells crafts at the local fairgrounds. She bought a 100-yard roll of ribbon. If she cuts the ribbon in $1\frac{1}{4}$-yard lengths, how many pieces could she get from the roll?

____

LESSON 7

Adding and Subtracting Fractions

When adding and subtracting fractions, follow these steps:

Step 1 Find a common denominator for the fractions.

Step 2 If the top fraction in a subtraction problem is less than the bottom fraction, rename the top whole number as a whole number and a fraction.

Step 3 Add or subtract the numerators of the fractions. Add or subtract the whole numbers.

Step 4 Reduce all answers.

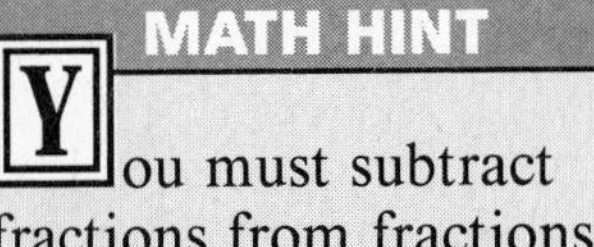

MATH HINT

You must subtract fractions from fractions and whole numbers from whole numbers.

Examples

A.

$$\begin{array}{r} 5\frac{3}{5} \\ +1\frac{1}{5} \\ \hline 6\frac{4}{5} \end{array}$$

B.

$$\begin{array}{rcl} 12\frac{4}{5} & = & 12\frac{12}{15} \\ +\ \frac{2}{3} & = & \frac{10}{15} \\ \hline & & 12\frac{22}{15} = 13\frac{7}{15} \end{array}$$

Find a common denominator.

Reduce.

C.

$$\begin{array}{rcl} 4 & = & 3\frac{3}{3} \\ -\ \frac{2}{3} & = & \frac{2}{3} \\ \hline & & 3\frac{1}{3} \end{array}$$

A fraction cannot be subtracted from a whole number until the whole number is renamed. Use the denominator of the fraction to rename 4 as $3\frac{3}{3}$.

D.

$$\begin{array}{rcccl} 15\frac{6}{15} & = & 15\frac{12}{30} & = & 14\frac{30}{30} + \frac{12}{30} = 14\frac{42}{30} \\ -\ 8\frac{7}{10} & = & 8\frac{21}{30} & = & 8\frac{21}{30} \\ \hline & & & & 6\frac{21}{30} = 6\frac{7}{10} \end{array}$$

Step 1 (finding the common denominator), Step 2 (renaming), Step 3 (subtracting), Step 4 (reducing)

Practice

Add or subtract. Reduce answers to lowest terms.

1. $\begin{array}{r} 16\frac{8}{11} \\ -\ 4\frac{3}{11} \\ \hline \end{array}$

2. $\begin{array}{r} 105\frac{1}{10} \\ +\ 82\frac{4}{10} \\ \hline \end{array}$

3. $\begin{array}{r} 16\frac{7}{10} \\ -\ 7\frac{1}{5} \\ \hline \end{array}$

4. $\begin{array}{r} 57\frac{1}{8} \\ +66\frac{2}{5} \\ \hline \end{array}$

5. $\begin{array}{r} 19\frac{1}{3} \\ -12\frac{7}{9} \\ \hline \end{array}$

6. $\begin{array}{r} 57\frac{1}{2} \\ 2\frac{1}{3} \\ +\ 9 \\ \hline \end{array}$

7. $\begin{array}{r} 7 \\ -\ \frac{3}{4} \\ \hline \end{array}$

8. $\begin{array}{r} 21\frac{3}{8} \\ -\ 9\frac{2}{3} \\ \hline \end{array}$

Problem Solving

Solve the following problems. Write your answers in lowest terms.

9. To wire the speakers to the stereo, John needs $24\frac{1}{6}$ feet of wire for the living room and $13\frac{1}{3}$ feet for the dining room. How much wire does he need in all? ____________

10. The Gonzalez's home is exactly $44\frac{1}{2}$ miles from their daughter's house. If they stop for gas after driving $21\frac{1}{8}$ miles, how much further do they have to drive to reach their daughter's house? ____________

11. The roll of yellow emergency tape was 110 feet long. The police used $97\frac{1}{2}$ feet of it to mark off the area for the spectators. How much was left? ____________

12. To pick up three friends, Jessica drove $4\frac{1}{4}$ miles to the first house, $6\frac{1}{3}$ miles to the second house, and $2\frac{1}{6}$ miles to the third house. How far did she drive in all? ____________

13. The Taylors bought $108\frac{1}{4}$ feet of fencing. They used $66\frac{3}{8}$ feet. How much do they have left? ____________

Percents

Percents are parts of 100. A percent can be written as a fraction or a decimal.

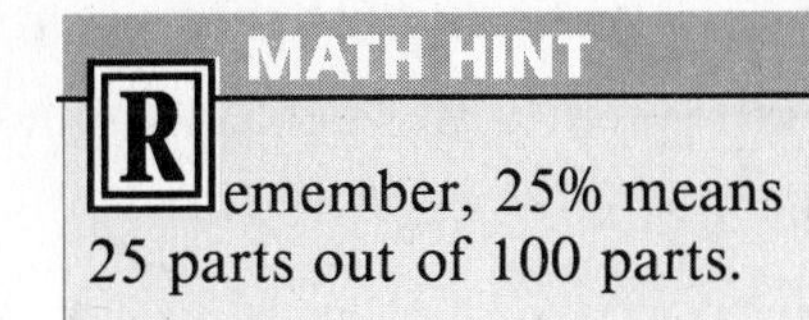

Examples

To change a percent to a decimal, follow these steps:

Step 1 Remove the percent sign (%).

Step 2 Multiply by .01.

Change the following percents to decimals.

A. 66%

Step 1 66

Step 2 $\times .01 = .66$

B. 200%

Step 1 200

Step 2 $\times .01 = 2.00$

To change a decimal to a percent, follow these steps:

Step 1 Multiply by 100.

Step 2 Add the percent sign.

Change the following decimals to percents.

C. .1

Step 1 $.1 \times 100$

Step 2 10%

D. 8

Step 1 8×100

Step 2 800%

To change a percent to a fraction, follow these steps:

Step 1 Remove the percent sign.

Step 2 Multiply by $\frac{1}{100}$.

Step 3 Cancel and reduce.

Change the following percents to fractions.

E. 40%

Step 1 $40 \times \frac{1}{100}$

Steps 2 and 3 $\overset{2}{\cancel{40}} \times \frac{1}{\underset{5}{\cancel{100}}} = \frac{2}{5}$

F. $33\frac{1}{3}\%$

Step 1 $33\frac{1}{3}$

Steps 2 and 3 $33\frac{1}{3} = \frac{100}{3}$

$\frac{\overset{1}{\cancel{100}}}{3} \times \frac{1}{\underset{1}{\cancel{100}}} = \frac{1}{3}$

To change a fraction to a percent, follow these steps:

Step 1 Multiply by $\frac{100}{1}$.

Step 2 Cancel and reduce. Make any remainders into fractions.

Step 3 Add the percent sign.

Change the following fractions to percents.

G. $\frac{1}{2}$

Steps 1 and 2 $\frac{1}{\cancel{2}_1} \times \frac{\cancel{100}^{50}}{1} = 50$

Step 3 50%

H. $\frac{2}{3}$

Steps 1 and 2 $\frac{2}{3} \times \frac{100}{1} = \frac{200}{3} = 66\frac{2}{3}$

Step 3 $66\frac{2}{3}\%$

Practice

Write the following percents as decimals.

1. 25% _____ **2.** 6% _____ **3.** 0.3% _____ **4.** 14% _____

5. 130% _____ **6.** 1.5% _____ **7.** 0.07% _____ **8.** 9% _____

Write the following percents as fractions. Reduce answers to lowest terms.

9. 20% _____ **10.** 17% _____ **11.** 4% _____ **12.** $5\frac{1}{2}\%$ _____

13. 96% _____ **14.** $66\frac{2}{3}\%$ _____ **15.** $\frac{1}{4}\%$ _____ **16.** 240% _____

Write the following decimals and fractions as percents.

17. 0.6 _____ **18.** $\frac{4}{10}$ _____ **19.** 0.01 _____ **20.** $\frac{5}{8}$ _____

21. $1\frac{1}{2}$ _____ **22.** 5 _____ **23.** $\frac{3}{5}$ _____ **24.** 0.009 _____

Problem Solving

Solve the following problems.

25. There was a $\frac{1}{6}$ decrease in the crime rate. Write the decrease as a percent. Round to the nearest whole percent. ________

26. A fruit juice drink contains 12% grapefruit juice. Write the amount as a fraction in lowest terms. ________

27. Five percent of Joe's pay is withheld for medical insurance. What is the decimal? ________

28. Of all the passengers, 0.04 of them had never flown before. What percent of the passengers has never flown before? ________

Comparing Fractions, Decimals, and Percents

Fractions, decimals, and percents are different ways to write equivalent values. To compare values, write them all as fractions or all as decimals.

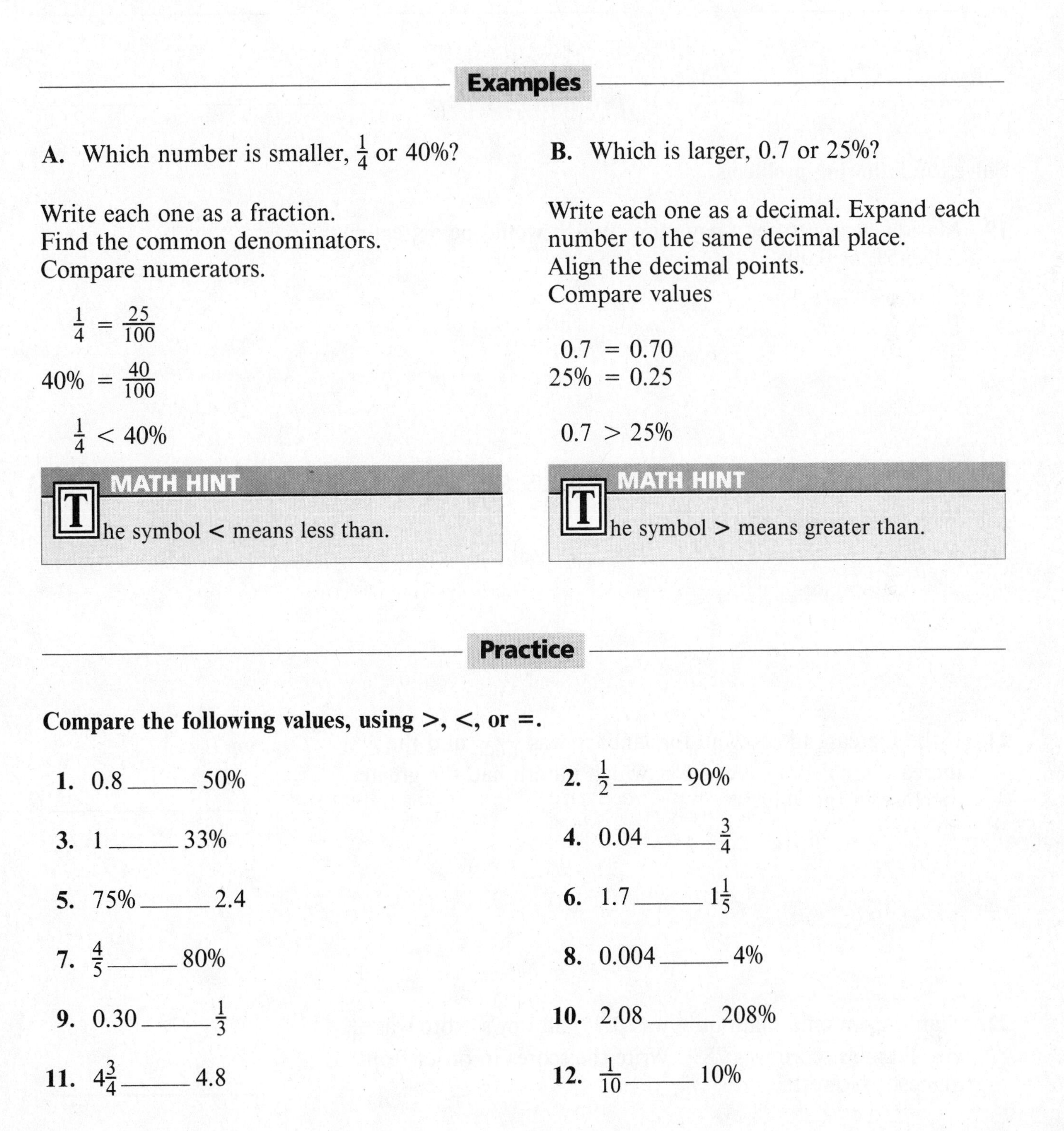

Examples

A. Which number is smaller, $\frac{1}{4}$ or 40%?

Write each one as a fraction.
Find the common denominators.
Compare numerators.

$$\frac{1}{4} = \frac{25}{100}$$

$$40\% = \frac{40}{100}$$

$$\frac{1}{4} < 40\%$$

MATH HINT

The symbol $<$ means less than.

B. Which is larger, 0.7 or 25%?

Write each one as a decimal. Expand each number to the same decimal place.
Align the decimal points.
Compare values

$$0.7 = 0.70$$
$$25\% = 0.25$$

$$0.7 > 25\%$$

MATH HINT

The symbol $>$ means greater than.

Practice

Compare the following values, using $>$, $<$, or $=$.

1. 0.8 _____ 50%

2. $\frac{1}{2}$ _____ 90%

3. 1 _____ 33%

4. 0.04 _____ $\frac{3}{4}$

5. 75% _____ 2.4

6. 1.7 _____ $1\frac{1}{5}$

7. $\frac{4}{5}$ _____ 80%

8. 0.004 _____ 4%

9. 0.30 _____ $\frac{1}{3}$

10. 2.08 _____ 208%

11. $4\frac{3}{4}$ _____ 4.8

12. $\frac{1}{10}$ _____ 10%

Write in order from the largest to the smallest.

13. $\frac{5}{8}$, 70%, 0.61 ______

14. 1.2%, $1\frac{1}{2}$, 0.6 ______

15. 0.045, 4%, $\frac{4}{10}$ ______

16. 0.6%, $\frac{3}{5}$, 0.01 ______

17. $\frac{3}{10}$, 33%, 1.3 ______

18. 150%, 1.8, $1\frac{1}{5}$ ______

Problem Solving

Solve the following problems.

19. Marie is going to get a pay raise. Which would be a greater raise, 5% or 0.005? ______

20. Matthew says that $\frac{11}{10{,}000}$ is less than 0.005. Steve says the two values are equal. Who is correct? ______

21. If the increase in snowfall for January was $\frac{3}{1{,}000}$ and the increase for February was 3%, which month had the greater increase in snowfall? ______

22. Alan's score on a math quiz was 0.75, Shirley's score was $\frac{36}{50}$, and Elaine's score was 78%. Write the scores in order from highest to lowest. ______

LESSON 10

Ratios and Proportions

A **ratio** shows the relationship of one number to another. A ratio can be written with a colon (:), the word **to**, or as a **fraction.**

MATH HINT

Other words such as *a, an, in, for, at,* and *per* are also used to indicate ratios.

Examples

A. At a banquet there are 10 chairs to a table. The ratio of chairs to tables can be written as follows:

10:1 or 10 to 1 or $\frac{10}{1}$

A **proportion** is a statement that two ratios are equal.

B. $\frac{3}{4} = \frac{9}{12}$ $\frac{8}{9} = \frac{64}{72}$

MATH HINT

All equal fractions are proportions.

The cross products of a proportion are equal.

C. $\frac{3}{5} = \frac{24}{40}$

$3 \times 40 = 5 \times 24$

$120 = 120$

To find the missing information in a proportion problem, follow these steps:

Step 1 Let the letter *n* stand for the missing number.

Step 2 Multiply the known cross products.

Step 3 Divide by the number that is left.

D. $\frac{4}{5} = \frac{?}{35}$

Step 1 $\frac{4}{5} = \frac{n}{35}$

Step 2 $4 \times 35 = 140$

Step 3 $\frac{140}{5} = n$

$n = 28$

Practice

Write the following ratios using a colon (:). Then write the ratios as fractions. Reduce all your answers.

1. 5 inches on a map to 50 miles ________ ________

2. 9 wheels to 3 tricycles ________ ________

3. 7 cars for 3 families ________ ________

4. 24 students per class ________ ________

Find the missing information in the following proportions.

5. $\frac{n}{5} = \frac{16}{20}$ ________

6. $\frac{2}{3} = \frac{n}{12}$ ________

7. $\frac{5}{8} = \frac{25}{n}$ ________

8. $\frac{22}{n} = \frac{11}{50}$ ________

9. $\frac{n}{3} = \frac{16}{24}$ ________

10. $\frac{3}{4} = \frac{n}{16}$ ________

11. $\frac{7}{9} = \frac{49}{n}$ ________

12. $\frac{12}{n} = \frac{18}{36}$ ________

13. $\frac{15}{n} = \frac{5}{15}$ ________

Problem Solving

Solve each problem using proportions.

14. One gallon of paint will cover 450 square feet. If the room is 2,700 square feet, how many gallons of paint are needed? ________

15. If doughnuts cost $5.76 a dozen, how much does one doughnut cost? ________

16. Horace gets a base run every four times at bat. If he goes to bat 160 times during the season, how many base runs is he likely to get? ________

17. At the grocery store, 4 bars of soap sell for $2.40. How much will 12 bars cost? ________

LESSON 11

Solving Percent Problems

Percent problems are a kind of proportion problem. They can be written in one of two forms.

60% of 200 is 120.   40 is 25% of 160.

All percent problems have three components:

The **percent** is the number just before the **%** sign.
The **whole** is the number that follows the word **of.**
The **part** is the number before or after the word **is.**

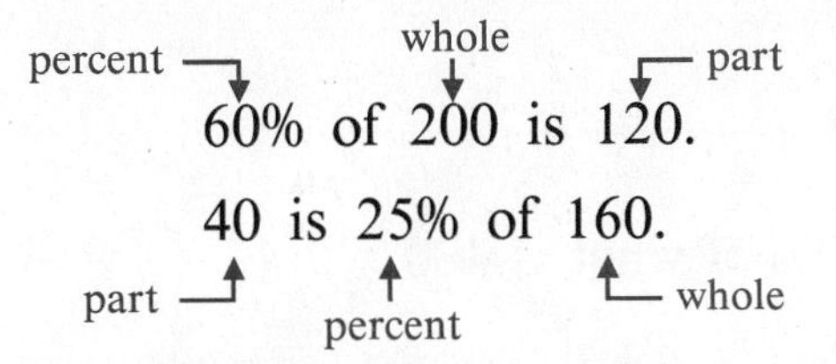

You can organize percent problems into grids like the one shown below.

part	percent
whole	100

MATH HINT

By using a grid, what you are really doing is setting up a proportion.

Example

To solve percent problems, follow these steps:

45 is what percent of 60?

Step 1 Place the known information on a grid.

Step 1

45 part	? percent
60 whole	100 100

The proportion is $\frac{45}{60} = \frac{n}{100}$

Step 2 Multiply the known diagonals.

Step 2 $45 \times 100 = 4{,}500$

Step 3 Divide by the number that is left.

Step 3 $\frac{4{,}500}{60} = 75$

Step 4 If the answer is a percent, add the percent sign.

Step 4 75%

45 is 75% of 60.

Practice

Solve the following percent problems.

1. 20% of 95 is what? __________

2. 53% of what is 1,855? __________

3. 49 is what percent of 700? __________

4. 85% of 220 is what? __________

5. 21 is what percent of 60? __________

6. 12% of what is 144? __________

7. 95% of 20 is what? __________

8. 200% of what is 400? __________

Problem Solving

Solve the following problems.

9. Seventy-five percent of the concert audience were under the age of 16. If there were 400 people in the audience, how many were under the age of 16? __________

10. Forty-two percent of the books on the shelf are fiction. The rest are nonfiction. If 21 books are fiction, how many books are on the shelf? __________

11. The number of salespeople in the department is 200% of what it was last year. If there were 12 salespeople in the department last year, how many are there now? __________

12. Nine of the fifty students in the class thought the assignment was too hard. What percent of the class thought the assignment was too hard? __________

13. Janelle spends 38% of her monthly salary on car payments and rent. If those expenses are $912, how much does she make each month? __________

Finding Simple Interest

Interest is the cost of borrowing money. Simple interest problems are like percent problems except interest problems always include time.

To find the interest, you need to know the **principal** (the amount of money being borrowed), the **rate** (the percent of the principal being paid per year as interest), and the **time** (the length of the loan expressed in years).

Part = whole × percent
Interest = principal × rate × time

MATH HINT

To solve simple interest problems, change the percent (rate) to a fraction or decimal.

To find simple interest, follow these steps:

Step 1 Draw a grid with interest as the part, rate as the percent, and principal as the whole.

Step 2 Multiply the known diagonals.

Step 3 Divide by the number that is left.

Step 4 Multiply the interest by the years (time) the money will be used.

Step 5 Write the answer in dollars and cents.

Example

What is the interest on $600 at a rate of 4% a year for 15 years?

Step 1

? interest	4 rate
$600 principal	100 100

Step 2 4 × $600

Step 3 $\frac{4 \times \overset{6}{\cancel{\$600}}}{\underset{1}{\cancel{100}}} = \$24$ interest for one year

Step 4 $24 × 15 years = $360

Step 5 $360 interest for 15 years at 4% on $600.

Practice

Find the amount of interest on the following.

1. $800 at 9% for 10 years.

______

2. $6,300 at 10% for 8 years and 6 months.

______

3. $75 at 3% for 1 year.

______

4. $1,500 at 18% for 9 months.

______

Problem Solving

Solve the following problems.

5. Lu Ann and Bill are getting a 4% home improvement loan for $7,562 for 3 years. How much interest will they pay? ______

6. Linda has a college loan for $2,000 at $6\frac{1}{2}$% interest. At the end of 2 years, how much interest will she pay? ______

7. Trent has $12,000 in a savings account earning 4% interest. If he keeps the money in the account for 12 years, how much interest will he receive? ______

8. What is the cost of a bank loan of $5,000 at 15% interest for 5 years and 3 months? ______

9. If Julie's department store credit card has a 19.5% interest rate and her balance is $1,650, how much interest will she pay in one year? ______

Problems 10–12 are related.

10. The Tylers have a 25-year mortgage loan for $60,000 with an interest rate of $8\frac{1}{2}$%. How much interest will they pay? ______

11. If the loan was for 15 years instead of 25, how much interest would they pay? ______

12. If the interest rate was 10%, how much interest would they pay? ______

LIFE SKILL

Installment Plan Buying

Installment plan buying means paying a certain percent of the purchase price as a down payment and agreeing to pay the rest in small amounts each month. The monthly amounts include an interest payment. This type of interest is called a **finance charge.**

Connie bought a couch on sale for $725. She put down a 10% deposit and paid $55 a month for a year. What total amount did she pay?

Deposit = $725 × 0.10	= $ 72.50
$55 a month for 12 months	= $660.00
Total cost of the couch	$732.50

Solve the following problems.

1. Yao wants to buy a color TV for $2,000. The store says he can put 15% down and pay $160 a month for 12 months. What is the total amount he will pay for the TV? ______________

2. Emily bought a winter coat for $400. She put 25% down and paid $80 a month for four months. What was the total cost of the coat? ______________

3. Horatio bought a computer for $2,800. He put 35% down and paid $165 a month for a year. What was the total cost of the computer? ______________

4. Consuela bought a dining room set for $1,800. She put 15% down and paid $90 a month for 18 months. What was the total cost of the dining room set? ______________

Percent of Increase or Decrease

Percent of change is used to compare the difference between two amounts. If the new amount is greater than the original amount, there is a **percent of increase.** If the new amount is less than the original amount, there is a **percent of decrease.**

To find the percent of increase or decrease, follow these steps:

Step 1 Find the difference between the original amount and the new amount.

Step 2 Divide the difference by the original amount.

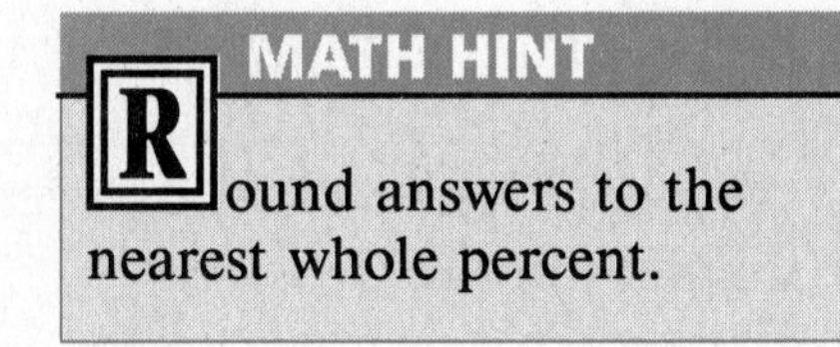

Examples

A. The Chus bought a car last year for \$9,750. This year the same model costs \$10,725. What is the percent of increase?

Step 1 \$10,725 – \$9,750 = \$975

Step 2 $\frac{\$975}{\$9750} = 0.1$

0.1 changed to a percent is 10%.

The percent of increase is 10%.

B. Rose bought a car 2 years ago for \$8,642. She sold it last week for \$5,650. What is the percent of decrease?

\$8,642 – \$5,650 = \$2,992

$\frac{\$2,992}{\$8,642} = 0.35$

0.35 changed to a percent is 35%.

The percent of decrease is 35%.

Practice

Determine whether each change is a percent of increase or a percent of decrease.

1. original price: \$17.86; new price: \$23.85

______

2. original cost: \$1,280; new cost: \$1,140

______

Problem Solving

Solve the following problems.

3. The weekly wages of a chef rose from $750 to $900 in 3 years. What was the percent of increase? __________

4. Electronic calendars decreased in price from $125 to $75 in the past year. What is the percent of decrease? __________

5. The price of a fabric dropped from $5.50 per yard to $4.30 per yard during a sale. What was the percent of decrease? __________

6. The price of pork changed from $2.40 a pound to $3.00 a pound. What is the percent of increase? __________

7. Henry bought a computer on sale for $2,000. It originally sold for $2,600. What was the percent of decrease? __________

8. Flora and John Sawyer bought a house for $60,000 seventeen years ago. They just sold it for $120,000. What was the percent of increase in the value of the house? __________

9. Casey bought a CD player last year for $355. This year, the CD player sells for $225. What is the percent of decrease in the selling price? __________

10. Darla's father bought a gold watch 50 years ago for $25. The watch is now worth $600. What is the percent of increase in the value of the watch? __________

LIFE SKILL

Working With Time Zones

The United States, including Hawaii and most of Alaska, is in five main time zones. From Eastern Standard Time on the east coast to Pacific Standard Time on the west coast, each time zone to the left, or the west, is one hour earlier. Hawaii and most of Alaska are two hours earlier than Pacific Standard Time because there is another time zone between Pacific Standard Time and Alaska Standard Time.

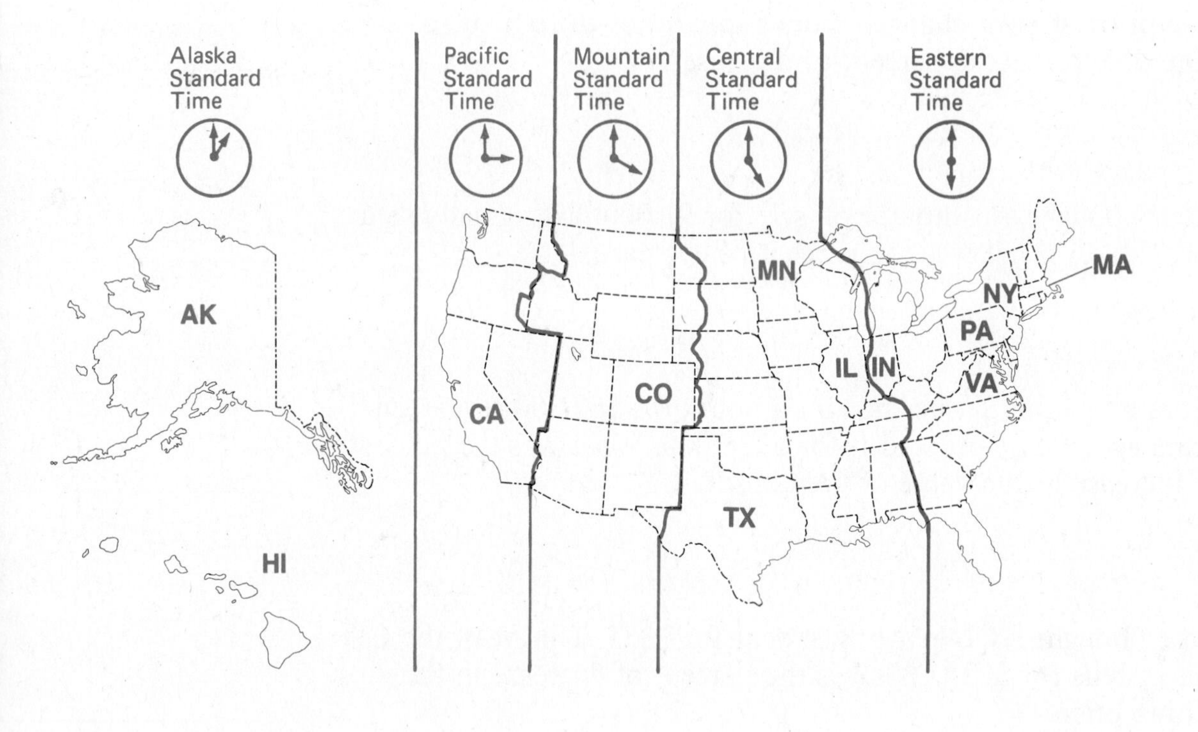

Answer these questions using the information on the map.

1. If it is 6 p.m. in New York, what time is it in Colorado? ______________

2. At 3:25 p.m. a person in Duluth, MN, called Philadelphia, PA. What time was it in Philadelphia? ______________

3. Fred took a charter bus from the Naval Station at Norfolk, VA, to San Diego, CA. How many hours did he set back his watch? ______________

4. Swan Lee drove 8 hours and 30 minutes from her home in Chicago, IL, to Bloomington, IN. If she left at 9:45 a.m., at what time did she arrive? ______________

5. Bonnie took a plane from Boston, MA to Houston, TX. The flight took 3 hours and 30 minutes. If she left at 5:50 p.m., at what time did she arrive? ______________

LESSON 14

Problem Solving With Decimals, Fractions, Percents, and Proportions

The following four steps can be used to solve word problems, including those involving decimals, fractions, percents, and proportions.

Step 1 Read the problem and underline the key words. These words will usually relate to some mathematics reasoning computation.

Step 2 Make a plan to solve the problem. Ask yourself, Should I add, subtract, multiply, divide, round, or compare? You may have to do more than one of these operations for the same problem.

Step 3 Find the solution. Use your math knowledge to find your answer.

Step 4 Check the answer. Ask yourself, Is this answer reasonable? Did you find what you were asked for?

When you begin with the first step, read the problem carefully. Look for clue words that will give you important information. For example, some problems will provide information about proportions.

Example

Justin can store 18 toy cars per box. How many boxes will he need to store 72 toy cars?

Step 1 The key word is **per.** The two things being compared are toy cars and boxes.

Step 2 You should solve using a proportion.

$$\frac{\text{toy cars}}{\text{boxes}} = \frac{18}{1} = \frac{72}{n}$$

Step 3 Find the solution by multiplying the known cross products, then divide by the number that is left.

$$\frac{1 \times 72}{18} = 4 \text{ boxes}$$

Step 4 Check the answer. Does it make sense that Justin will need 4 boxes to store 72 cars? Yes, the answer is reasonable.

Practice

Solve the following problems using the steps to solve word problems.

1. Mitch has a board $12\frac{1}{2}$ feet long. He needs to cut it into 6-inch pieces. How many pieces will he have? ________

2. Alexis paid $24.95 for a birthday gift, $21.50 for a cake, and $6.25 for decorations. How much did she spend in all? ________

3. A bakery has 32 blueberry bagels. If there are 160 bagels in all, what percent of them are blueberry? ________

4. Arthur uses 4 crates to pack 120 ears of corn. If there are 450 ears of corn, how many crates will he need? ________

5. Seventy-six percent of the students drive to school. If 1,140 students drive to school, how many students are there in all? ________

UNIT

1

Posttest

Solve the following.

1. $\begin{array}{r} 18.402 \\ +\ 7.93 \\ \hline \end{array}$

2. $\begin{array}{r} 416.85 \\ -\ 8.276 \\ \hline \end{array}$

3. $\begin{array}{r} 14.05 \\ \times\ 1.4 \\ \hline \end{array}$

4. $0.26\overline{)1.092}$

5. $\begin{array}{r} 17\frac{2}{3} \\ +\ 56\frac{5}{8} \\ \hline \end{array}$

6. $\begin{array}{r} 109\frac{1}{6} \\ -\ 42\frac{3}{8} \\ \hline \end{array}$

7. $5\frac{2}{9} \times \frac{3}{4}$ ______________

8. $7\frac{1}{3} \div 1\frac{4}{7}$ ______________

Write the equivalent of each of the following.

	Fraction	Decimal	Percent
9.	______	0.05	______
10.	$\frac{4}{5}$	______	______
11.	______	______	14%
12.	______	3	______
13.	$\frac{5}{8}$	______	______
14.	______	______	1.5%

Compare the following, using <, >, or =.

15. 1.006 ______ .1

16. $\frac{3}{10}$ ______ 0.003

17. 1% ______ 0.1

18. $4\frac{1}{5}$ ______ 4.2

19. 37.5% ______ $\frac{3}{8}$

20. $\frac{9}{100}$ ______ 0.009

Problem Solving

Solve the following problems.

21. In the morning, there were 500 cans in the soup display. During the day, 12% of the display was sold. How many soup cans were remaining? ______________

22. At the park, the ratio of pine trees to shade trees is 4:3. If there are 56 pine trees, how many shade trees are there? ______________

23. Jean has $982 in a 3-month savings certificate at 6% interest. What will be the interest on the certificate? ______________

24. Werner bought a new car on sale for $9,600. The car originally cost $12,000. What was the percent of decrease? ______________

25. The number of students enrolled in GED has increased this month. If there are 63 students enrolled now and that is 105% of the number enrolled last month, how many were enrolled last month? ______________

Measurement

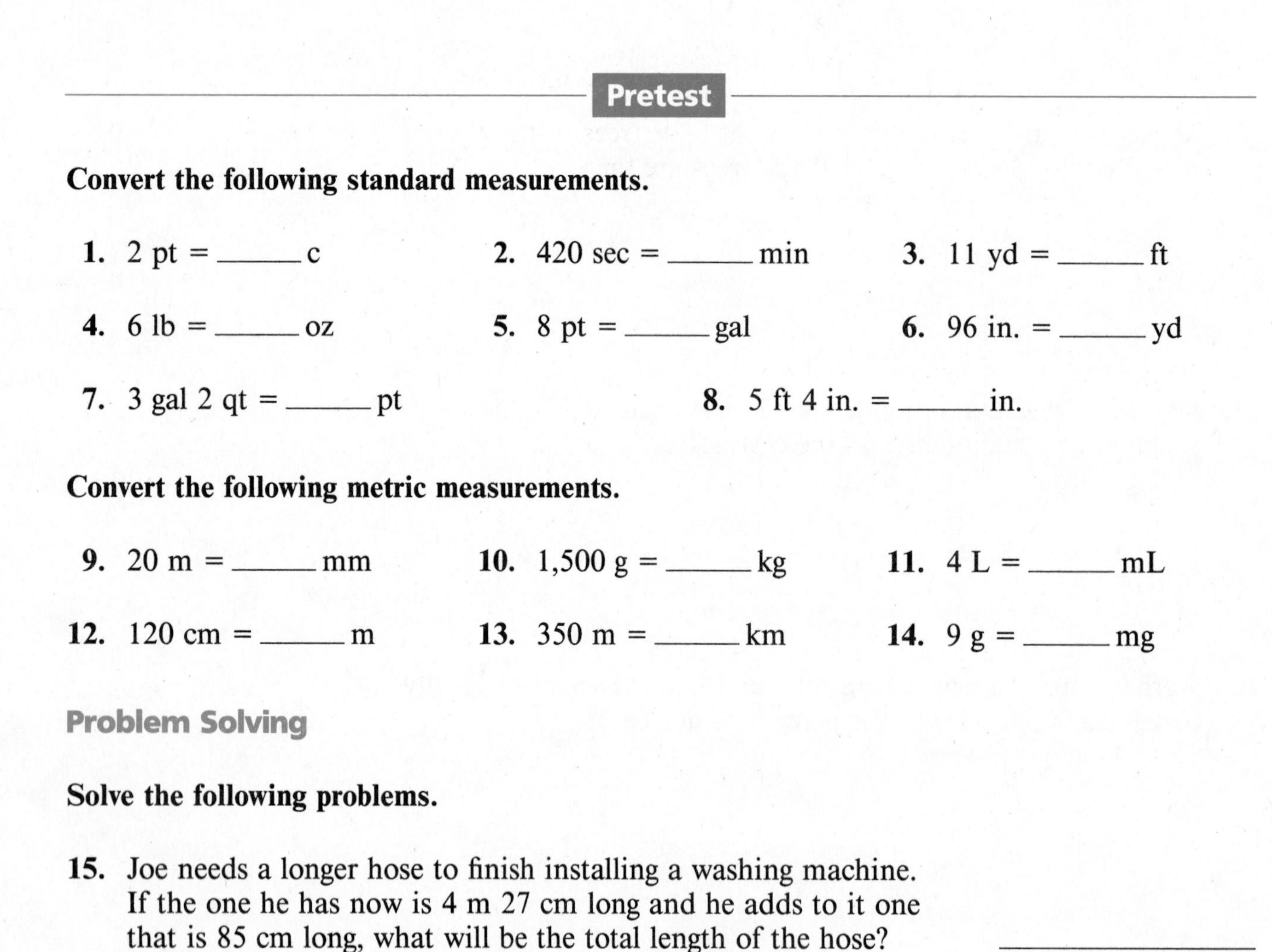

Pretest

Convert the following standard measurements.

1. 2 pt = ______ c

2. 420 sec = ______ min

3. 11 yd = ______ ft

4. 6 lb = ______ oz

5. 8 pt = ______ gal

6. 96 in. = ______ yd

7. 3 gal 2 qt = ______ pt

8. 5 ft 4 in. = ______ in.

Convert the following metric measurements.

9. 20 m = ______ mm

10. 1,500 g = ______ kg

11. 4 L = ______ mL

12. 120 cm = ______ m

13. 350 m = ______ km

14. 9 g = ______ mg

Problem Solving

Solve the following problems.

15. Joe needs a longer hose to finish installing a washing machine. If the one he has now is 4 m 27 cm long and he adds to it one that is 85 cm long, what will be the total length of the hose? ______________

16. Sandy spilled 7 ounces of sugar from a container holding 4 pounds 3 ounces. How much did she have left? ______________

17. Three people are planning a fishing trip to the lake. It takes 12 hours 45 minutes to get to the lake. If each drives an equal number of hours, how long will each person drive? __________

18. How many liters of soup does Juanita need to serve 5 people 250 mL each? __________

19. Karen plans on making 4 gallons of tomato sauce. How many jars will she need if each jar holds 1 quart? __________

20. To make the chemical solution needed to develop film, Anita stirs 50 mL of developer into 100 mL of warm water and then adds water until she has 1 liter. How much more water must she add? __________

Problems 21 and 22 are related.

21. AAA Hardware Store is selling insulating stripping for \$0.60 per meter. BBB Home Improvement Center is selling the same stripping for \$11.00 for 20 meters. Which store has the better buy? __________

22. At AAA Hardware Store, the sale on paint is \$9.95 per gallon. BBB Home Improvement Center is selling the same paint for \$3.05 per quart. Which store has the better buy? __________

Introduction to Measurement

Two types of measurement systems are used in the world—the standard and the metric systems. The pictures compare similar metric and standard measurement units. Look at them carefully.

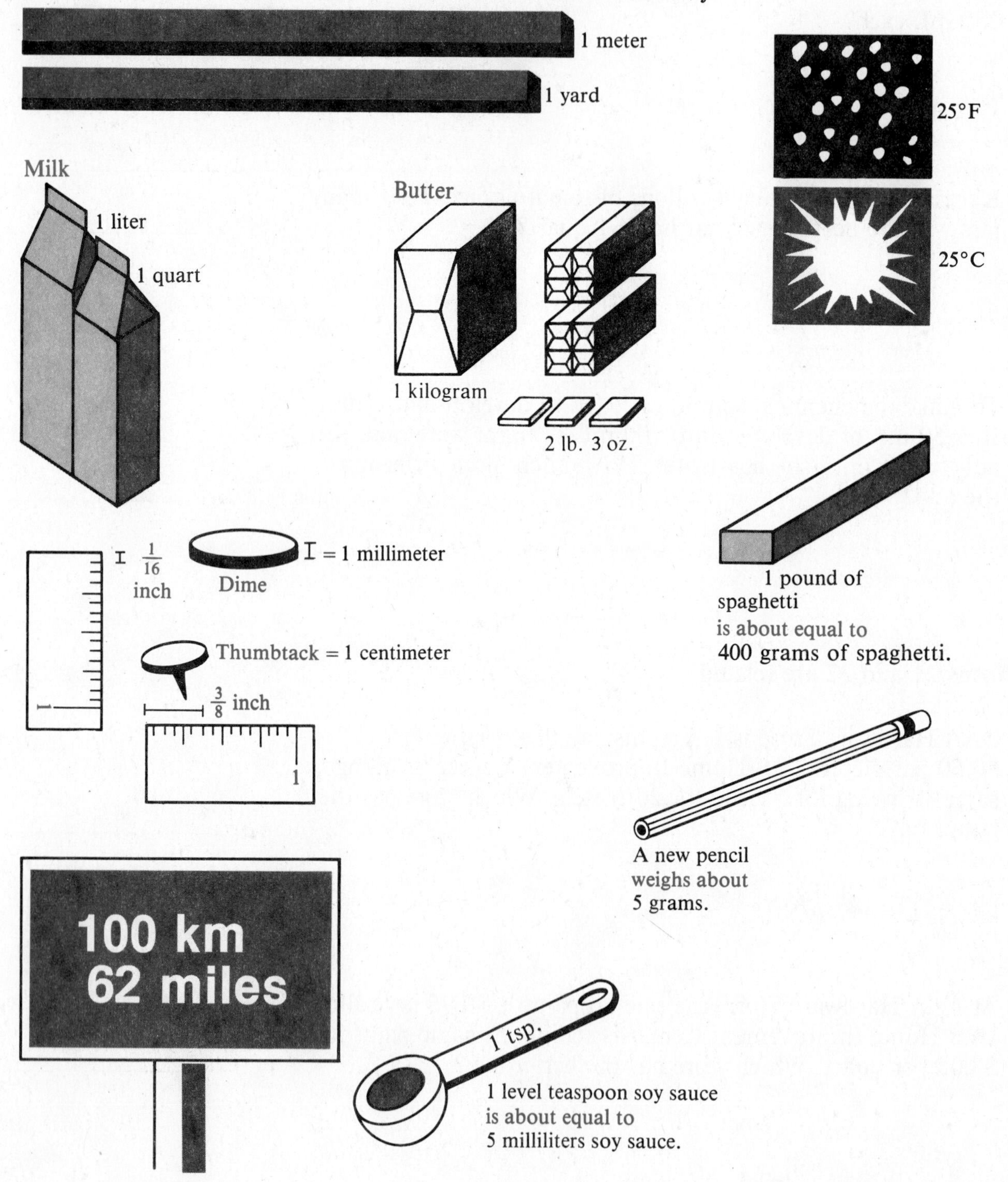

Practice

Use the drawings on page 38 and your own experience to answer the following questions. Decide what kind of measure is being used. Then put the letter of the correct measure in the blank.

W Weight measure
V Volume measure
L Length measure

1. grams _____	**2.** yards _____	**3.** kilometers _____
4. gallons _____	**5.** kilograms _____	**6.** feet _____
7. inches _____	**8.** cups _____	**9.** ounces _____
10. centimeters _____	**11.** meters _____	**12.** liters _____
13. millimeters _____	**14.** pints _____	**15.** milligrams _____
16. pounds _____	**17.** tons _____	**18.** quarts _____

Match the metric unit to a familiar amount. Put the letter of the correct answer in the blank.

19. 100 kilometers	_____	**A.** about 1 yard
20. 1 liter	_____	**B.** about a teaspoon
21. 5 grams	_____	**C.** the thickness of a dime
22. 25°C	_____	**D.** a little more than 2 pounds
23. 1 meter	_____	**E.** the width of a thumbtack
24. a kilogram	_____	**F.** about 60 miles
25. 1 millimeter	_____	**G.** about the weight of a new pencil
26. 1 centimeter	_____	**H.** about 1 pound
27. 400 grams	_____	**I.** the temperature of a summer day
28. 5 milliliters	_____	**J.** about a quart

Converting Among Standard Measurements

The type of measurement system commonly used in the United States is the standard measurement system. Some of these measurements and their equivalents are listed below.

Volume

1 gallon (gal) = 4 quarts (qt)
1 quart = 2 pints (pt)
1 pint = 2 cups (c)

Time

1 week = 7 days
1 day = 24 hours (hr)
1 hour = 60 minutes (min)
1 minute = 60 seconds (sec)

Length

1 mile (mi) = 1,760 yards (yd) or 5,280 feet (ft)
1 yard = 3 feet
1 foot = 12 inches (in.)

Weight

1 ton (t) = 2,000 pounds (lb)
1 pound = 16 ounces (oz)

Example

How many gallons are in 64 quarts?

To convert from one measure to another, create a proportion using the relationship of one of the measures to the other.

Let n = the number of gallons in 64 quarts.

$$\frac{1 \text{ gallon}}{4 \text{ quarts}} = \frac{n \text{ gallons}}{64 \text{ quarts}}$$

Solve for n. Multiply the diagonals (cross-products) of the known numbers and divide by the number that is left.

$$n = \frac{1 \times \overset{16}{\cancel{64}}}{\underset{1}{\cancel{4}}} = 16 \text{ gallons}$$

There are 16 gallons in 64 quarts.

Practice

Convert the following measurements.

1. How many quarts are in 7 gallons?

____

2. How many inches are in 4 yards?

____

3. How many days are in 72 hours?

____

4. How many ounces are in 3 pounds?

____

5. How many minutes are in 300 seconds?

____

6. How many pints are in 6 quarts?

____

7. How many yards are in 20 miles?

____

8. How many minutes are in a day?

____

9. How many inches are in 10 feet?

____

10. How many pounds are in 16 tons?

____

11. How many quarts are in 40 gallons?

____

12. How many seconds are in 1 hour?

____

Problem Solving

Solve the following problems.

13. Jason is planning a schedule for the next 5 weeks. His plant operates 24 hours a day, 7 days a week. How many hours must he schedule?

____

14. Jose is ordering pipe to lay along the road. The roadwork is 2 miles long. The pipe is ordered in yards. How many yards must he order?

____

15. Francine must order dairy mix for the ice cream machine. The machine holds 8 gallons. How many quarts of mix must she order?

____

16. Manuel must run fencing along 600 feet of his property. How many yards of fencing must he order?

____

17. The truck is carrying 7,000 pounds of goods. If its cargo is more than 3 tons, the driver will have to pay a fine. Does this driver have to pay a fine?

____

18. The machine in the workshop must be oiled every 72 hours of operation. It has operated 24 hours a day for 2 days. Does it need oil?

____

LESSON 17

Writing Fractions as Standard Measures

To write a fraction as a standard measure using only whole units, convert the fraction to a smaller unit of measure.

Examples

A. Write $2\frac{3}{4}$ feet as a measure using whole units.

$2\frac{3}{4}$ feet means 2 feet and $\frac{3}{4}$ of a foot.
Convert $\frac{3}{4}$ of a foot to inches.

$\frac{3}{4} = \frac{n}{12}$ Set up a proportion.

$n = \frac{3 \times 12}{4}$ Multiply the known diagonals and divide by the number that is left.

$= 9$ inches

So, $2\frac{3}{4}$ feet $=$ 2 feet 9 inches.

B. Write $\frac{1}{2}$ gallon as a measure using whole units.

Convert $\frac{1}{2}$ gallon to quarts.

$\frac{1}{2} = \frac{n}{4}$ Set up a proportion.

$n = \frac{1 \times 4}{2}$ Multiply the known diagonals and divide by the number that is left.

$= 2$ quarts

So, $\frac{1}{2}$ gallon $=$ 2 quarts.

Practice

Convert the following fractions to standard measures using whole units.

1. $4\frac{1}{2}$ gallons _____ gallons _____ quarts

2. $2\frac{1}{2}$ pints _____ pints _____ cup

3. $20\frac{3}{4}$ pounds _____ pounds _____ ounces

4. $2\frac{1}{4}$ tons _____ tons _____ pounds

5. $6\frac{1}{3}$ yards _____ yards _____ foot

6. $2\frac{1}{2}$ quarts _____ quarts _____ pint

7. $3\frac{1}{4}$ hours _____ hours _____ minutes

8. $6\frac{1}{3}$ days _____ days _____ hours

9. $2\frac{1}{4}$ feet _____ feet _____ inches

10. $5\frac{1}{2}$ minutes _____ minutes _____ seconds

11. $\frac{1}{4}$ pound ______ ounces

12. $\frac{1}{2}$ ton ______ pounds

13. $\frac{1}{4}$ gallon ______ quart

14. $\frac{1}{10}$ ton ______ pounds

Problem Solving

Solve the following problems.

15. Anita has $5\frac{2}{3}$ yards of velvet ribbon in her sewing box. Write the amount of ribbon using whole units.

______

16. There are $2\frac{1}{2}$ quarts of homemade strawberry jam left in the cupboard. Write the amount of jam left using quarts and pints. ______

17. Abe worked on his model ship for $5\frac{3}{4}$ hours. Write the amount of time he worked using hours and minutes. ______

18. If a bowl of chocolate fudge weighs $5\frac{1}{4}$ pounds, then it weighs 5 pounds and ______ ounces. ______

LIFE SKILL

Calculating Overtime

Some companies pay people extra when they work more than a regular 40-hour work week. The overtime rate may be $1\frac{1}{2}$ times the regular hourly rate. This is called **time and a half.** It means one and one-half times the normal salary will be paid for every hour worked over 40 hours.

Julio is paid $12 an hour for a regular 40-hour week at the candy factory. He receives time and a half for the hours he works over 40. If he worked 43 hours this week, what is his total pay?

40 hours × $12	= $480	regular pay
3 hours × $12 × $1\frac{1}{2}$	= $ 54	overtime pay
	$534	total salary

Determine the number of hours worked and the weekly salaries for the following people. They receive time and a half for every hour over 40.

1. Hank is paid $14.40 an hour.

Monday	7 hr 30 min
Tuesday	9 hr
Wednesday	7 hr 15 min
Thursday	8 hr 45 min
Friday	9 hr 30 min

Total hours ____________

Weekly salary ____________

2. Caroline is paid $9 an hour.

Monday	8 hr 30 min
Tuesday	7 hr 45 min
Wednesday	10 hr 15 min
Thursday	9 hr 30 min
Friday	8 hr

Total hours ____________

Weekly salary ____________

LESSON 18

Writing Standard Measures as Fractions

To write a standard measure as a fraction, convert the measure to a larger unit.

Examples

A. Write 4 ounces as a fraction of a pound.

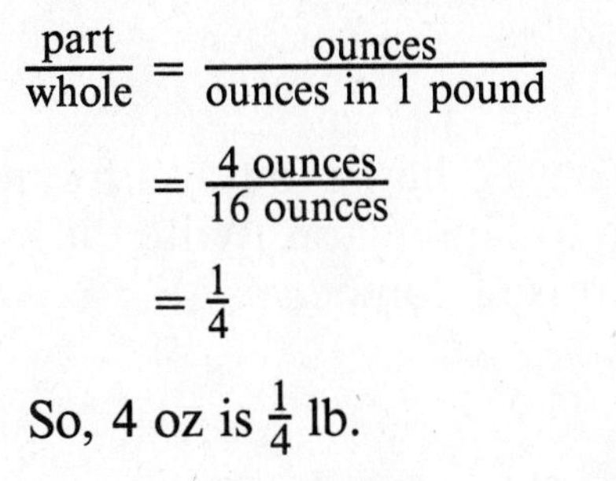

$$\frac{\text{part}}{\text{whole}} = \frac{\text{ounces}}{\text{ounces in 1 pound}}$$

$$= \frac{4 \text{ ounces}}{16 \text{ ounces}}$$

$$= \frac{1}{4}$$

So, 4 oz is $\frac{1}{4}$ lb.

B. Write 2 days 8 hours as a mixed number.

$$\frac{\text{part}}{\text{whole}} = \frac{\text{hours}}{\text{hours in 1 day}}$$

$$= \frac{8 \text{ hours}}{24 \text{ hours}}$$

$$= \frac{1}{3}$$

So, 2 days 8 hours is $2\frac{1}{3}$ days.

Practice

Write the following measurements as fractions or mixed numbers. Reduce all answers.

1. What fraction of a gallon is 3 quarts?

____

2. Write 3 yards 18 inches as a mixed number.

____

3. Write 2 feet 11 inches as a mixed number.

____

4. What fraction of a minute is 50 seconds?

____

5. What part of a ton is 100 pounds?

____

6. Write 12 miles 1,320 feet as a mixed number.

____

7. Write 2 hours 20 minutes as a mixed number.

______

8. What part of a pound is 12 ounces?

______

9. Write 5 pounds 4 ounces as a mixed number.

______

10. What part of a foot is 4 inches?

______

Problem Solving

Solve the following problems.

11. Stan used his electric saw to shorten the board to 3 feet 9 inches. Write the new length as a mixed number.

______

12. It takes 2 days 6 hours to travel from Gainsburg to Shillington. Write the time as a mixed number.

______

13. A small bird weighed 4 ounces. What fraction of a pound did the bird weigh?

______

14. The recipe called for 3 pints of cream to make the dessert. What fraction of a gallon was used?

______

LESSON 19

Working With Standard Measures

Examples

To add or subtract standard measures, add or subtract the like units and reduce. In subtraction problems, rename the top units if necessary.

A.
$$\begin{array}{r} 3\text{ lb } \ 4\text{ oz} \\ +\ 4\text{ lb } 12\text{ oz} \\ \hline 7\text{ lb } 16\text{ oz} \end{array} = 7\text{ lb} + 1\text{ lb} = 8\text{ lb}$$

B.
$$\begin{array}{rcr} 8\text{ hr} & = & 7\text{ hr } 60\text{ min} \\ -\ 3\text{ hr } 30\text{ min} & = & 3\text{ hr } 30\text{ min} \\ \hline & & 4\text{ hr } 30\text{ min} \end{array}$$

To multiply standard measures, multiply each unit and reduce.

C.
$$\begin{array}{r} 4\text{ ft } 9\text{ in.} \\ \times \quad 6 \\ \hline 24\text{ ft } 54\text{ in.} \end{array} = 24\text{ ft} + \frac{54}{12}\text{ ft}$$
$$= 24\text{ ft} + 4\tfrac{1}{2}\text{ ft}$$
$$= 28\tfrac{1}{2}\text{ ft}$$

To divide standard measures, change the measure to a mixed number and divide as usual.

D. 4 gal 2 qt ÷ 2

$$4\text{ gal } 2\text{ qt} = 4\tfrac{1}{2}\text{ gal} = \tfrac{9}{2}\text{ gal}$$
$$\tfrac{9}{2} \div 2 = \tfrac{9}{2} \times \tfrac{1}{2} = \tfrac{9}{4} = 2\tfrac{1}{4}\text{ gal}$$

Practice

Solve the following measurement problems. Reduce answers to lowest terms.

1. 2 qt 1 pt + 1 pt ________

2. 2 mi − 2,640 ft ________

3. 1 week 4 days × 5 ________

4. 1 hr 45 min ÷ 7 ________

5. 4 lb 6 oz + 1 lb 8 oz ________

6. 2 ft 4 in. ÷ 14 ________

7.
$$\begin{array}{r} 3\text{ yards } 24\text{ inches} \\ +\ 1\text{ yard } \ 12\text{ inches} \\ \hline \end{array}$$

8.
$$\begin{array}{r} 6\text{ yards } 1\text{ foot} \\ -\ 3\text{ yards } 2\text{ feet} \\ \hline \end{array}$$

9. 4 feet 6 inches
$\times$ 8

10. 7 gallons 1 quart
$-$ 2 gallons 3 quarts

Problem Solving

Solve the following problems.

11. Sargon's Quarry must deliver 2 tons 300 pounds of gravel to one building site and 1,700 pounds to another. What is the total amount of gravel to be delivered? ____________

12. Lori Giles uses 4 gallons 3 quarts of insect spray every time she dusts her crops. Since she sprays 5 times a year, how much insect spray will she need? ____________

13. It is 20 miles from the bridge to the cabin. One mile and 2,640 feet of this is a gravel road. The rest is paved. How long is the paved road? ____________

14. Three people volunteered to work at the blood drive. Each one worked at the booth for 2 hours 30 minutes. If they worked one after another, how many hours was the booth open? ____________

15. The coffee pot at the restaurant holds 10 gallons. The thermal coffee jugs hold 8 cups each. How many jugs can be filled from one pot of coffee? ____________

LIFE SKILL

Figuring Amount and Cost of Home Weatherproofing

"Our apartment needs some work before winter. Don't forget, the lease we signed says we pay for heat," said Diana. "You're right," said Thaddeus. "If we weatherize it now, we'll save some money on heating bills."

1. "We have 3 large windows. The distance around each one is $5\frac{1}{4}$ yards. We'll need ______ yards of weather stripping for them." ____________

2. "We'll also need $6\frac{3}{4}$ yards of stripping for the door. That's a total of ______ yards of weatherstripping." ____________

   "At $.40 a yard, that will cost ______." ____________

3. "If we buy clear plastic to cover the windows, we will cut down on heat loss. The plastic is wide enough to cover the width, so we only need 6 yards to cover the length. Add 6 inches for each of the 3 windows for overlap. That's a total of ______ yards of plastic." ____________

   "At $2.50 a yard, that's $______. It seems expensive, but if we're careful we can use it for more than one year." ____________

4. "Fifteen ounces of caulking compound will cover about 90 feet. I think we're going to need to cover about 3 times that much. So, we'd better get ______ ounces of caulk." ____________

LIFE SKILL

Reading a 24-Hour Clock

Twenty-four hour clocks are used in many places around the world. The military and some companies use a 24-hour clock in assigning a work schedule. This prevents people from confusing daytime hours with nighttime hours.

The 24-hour clock uses four digits. The first two digits refer to the number of hours that have passed since the beginning of the day. The second two digits refer to the number of minutes after the hour.

The 24-hour clock starts immediately after midnight counting time to 2400 hours. Beginning at 1 o'clock in the morning, the time is called 0100 hours (pronounced-*oh one hundred hours*).

Digital Clock	**24-Hour Clock**
2:45 in the morning	0245 hours
10:30 a.m.	1030 hours
12:59 p.m.	1259 hours

At 1 o'clock in the afternoon, 1200 is added to 0100. The time is 1300 hours. This means that 13 hours have passed since the start of the day.

Digital Clock	**24-Hour Clock**
1:00 p.m.	0100 + 1200 = 1300 hours
4:45 p.m.	0445 + 1200 = 1645 hours
5:00 in the afternoon	0500 + 1200 = 1700 hours
6:30 in the evening	0630 + 1200 = 1830 hours

Change the following standard times to 24-hour clock time.

1. Two o'clock in the afternoon __________

2. Six o'clock in the evening __________

3. Midnight __________

4. Fifteen minutes after 7 in the morning __________

5. Two in the morning __________

6. 10:20 a.m. __________

7. Eleven o'clock in the evening __________

8. 8:15 p.m. __________

9. 9:45 p.m. __________

10. Ten o'clock in the morning __________

Change the following 24-hour clock times into standard time. Remember to include a.m. or p.m.

11. 0542 hours __________

12. 1520 hours __________

13. 1200 hours __________

14. 1840 hours __________

15. 2340 hours __________

16. 0045 hours __________

17. 0100 hours __________

18. 1743 hours __________

19. 2215 hours __________

20. 0920 hours __________

LESSON 20

Calculating the Unit Price

Finding the unit price is helpful in determining which size of a product is the better buy based solely on price.

The unit price of an item is its cost per unit of measure, such as dollars per pound or cents per box. It is found by dividing the price by the number of units or the number of items.

$$\text{unit price} = \frac{\text{price}}{\text{number of units or items}}$$

Examples

A. Liza purchased a 16-ounce jar of salsa for $2.39 and 6 boxes of pudding for $3.30. What is the unit price of the items to the nearest tenth of a cent?

salsa: $\frac{\$2.39}{16 \text{ oz}} = \frac{0.149}{1 \text{ oz}}$

$= \$0.149$ per ounce

pudding: $\frac{\$3.30}{6 \text{ boxes}} = \frac{0.55}{1 \text{ box}}$

$= \$0.55$ per box

B. Cider is sold for $1.52 a gallon or $0.40 a quart. Which is the better buy?

1 gallon = 4 quarts

$\frac{\$1.52}{4 \text{ quarts}} = \frac{.38}{1 \text{ quart}} = \$0.38$ per quart

$\$0.38 < \$0.40$, so the gallon of cider is the better buy.

Practice

Find the unit price to the nearest tenth of a cent.

1. 2 pounds of bacon for $2.96

__________ per pound

2. canned tomatoes, $0.70 for 28 ounces

__________ per ounce

3. orange juice, $1.89 a gallon

__________ per quart

4. 3 video cassettes for $11.70

__________ per video cassette

5. cream, $1.20 a pint

__________ per cup

6. 36 tea bags for $3.99

__________ per tea bag

Problem Solving

Solve the following problems.

7. Paint thinner at Harold's Hardware is $0.89 per liter. It costs $0.65 for 800 mL at Bob's Supplies. Which store has the better buy? ____________

8. Trevor's favorite brand of nacho chips is on sale at Country Mart. The 28-ounce bag is $1.99 and the 36-ounce bag is $2.30. Which size is the better buy? ____________

9. Juliette is buying 2 cases of cola for $11.98. If each case has 24 cans, how much does each can cost? Round to the nearest hundredth of a cent. ____________

10. Sunshine Foods is selling 10-pound bags of potatoes for $2.45. Mark's Foods is selling 5-pound bags of potatoes for $1.20. Which store has the better buy? ____________

11. Fran's Market sells eggnog during the holidays for $2.10 a quart or $1.15 a pint. Which is the better buy? ____________

Reading a Thermometer

A thermometer measures temperature in degrees. The symbol for degree is °.

The two most common scales for measuring temperature are the Celsius and Fahrenheit scales. People living in countries using the metric system and scientists everywhere use the Celsius scale. Most people in the United States use the Fahrenheit scale.

Look at Diagram A. Many thermometers have both scales drawn on them. The left side of the thermometer is marked °C for Celsius and the right side is marked °F for Fahrenheit.

Read either scale of the thermometer like a number line. Look at the indicator letter and read across to the nearest number on the thermometer's scale. You will have to estimate the numbers that are not written.

A thermometer that you use to take a person's temperature usually ranges from 94°F to 106°F. Look at Diagram B. A number is written next to every two degrees and a small vertical line marks every two-tenths, or 0.2, of a degree.

Diagram A

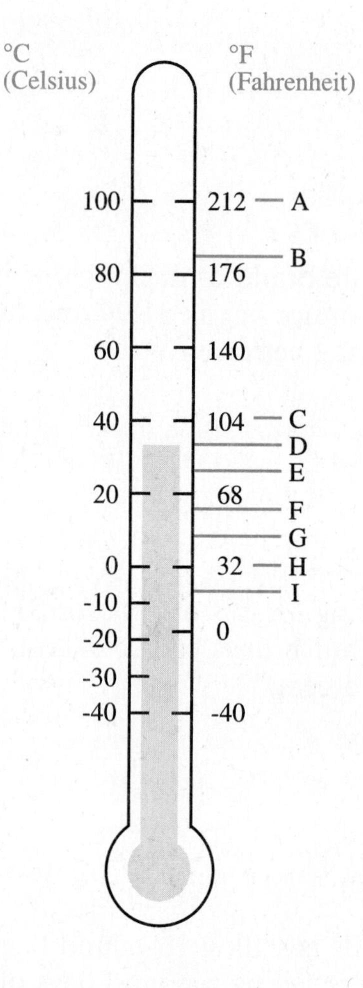

MATH HINT

An arrow points to 98.6° on the thermometer because that is the normal body temperature.

Diagram B

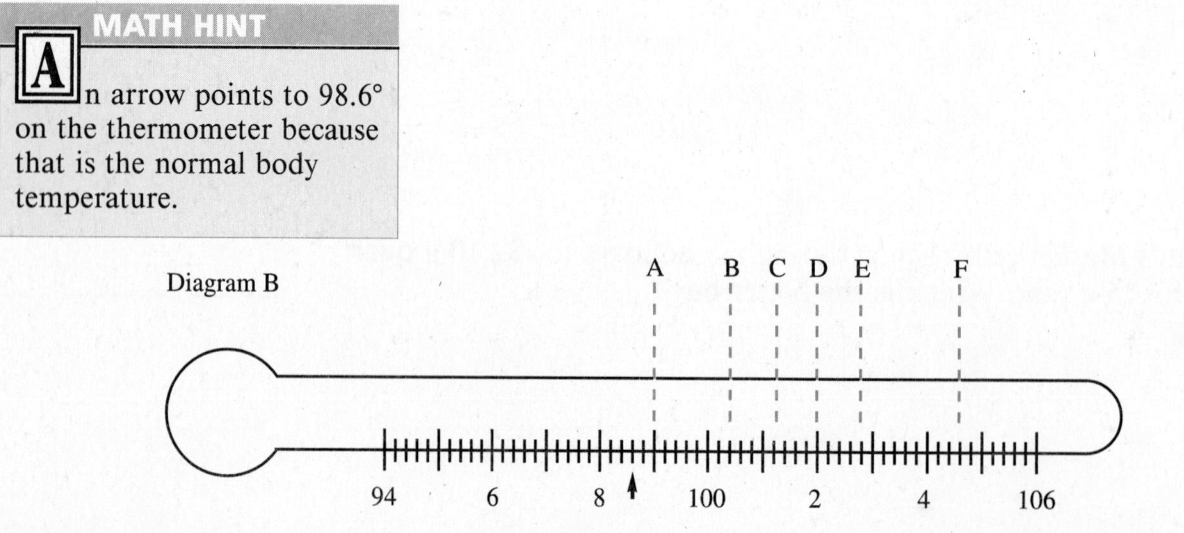

Examples

A. Look at Diagram A. What letter indicates a normal body temperature of 98.6°F or 37°C?

The letter D is closest to normal body temperature.

B. Refer to Diagram B. The letter D points to what temperature?

The letter D points to the line marked "2" which represents 102°F.

C. In Diagram B, the letter B points to what temperature?

The letter B is 2 marks past 100°F. Each mark represents 0.2 of a degree.

$100 + 0.2 + 0.2 = 100.4°F$

Practice

Refer to Diagram A. Write the letter on the thermometer that indicates each temperature listed below.

1. 104°F = 40°C ______

2. 41°F = 5°C ______

3. 32°F = 0°C ______

4. 185°F = 85°C ______

5. 25.6°F = −3.4°C ______

6. 212°F = 100°C ______

Refer to Diagram B. Find the temperatures for each of the letters marked on the thermometer.

7. E ______ **8.** A ______ **9.** C ______ **10.** F ______

Problem Solving

Solve the following problems.

11. If a person's temperature is 102.4°F, how many degrees is it above normal? ______

12. A baby's temperature is 103.4°F. If his temperature rises 1.6° more, he has to go see a doctor. At what temperature does the baby have to see a doctor? ______

LIFE SKILL

Rehabilitating a House

“This house is sound,” said Annie. “Even though it looks run-down, I know we can fix it up.”

“Well, the first thing we have to work on is the kitchen,” said Lew. “There are walls that need replacing and rewiring jobs to do.”

Solve the following problems.

1. “It’s 48 feet 9 inches around the room. We need matching baseboard and ceiling molding, so we’ll need a total of _____ feet.” ____________

2. “Don’t forget the two doors to the kitchen. They are each 2 feet 6 inches wide, so we only need _____ feet of molding,” said Annie. ____________

3. “The outside kitchen door needs weather stripping. It’s 2 feet 6 inches wide and $7\frac{1}{2}$ feet high. We’ll need _____ feet of stripping to go around,” Lew calculated. ____________

4. “My sister said we could have that oak board for shelves,” said Annie. “It’s 12 feet 8 inches long. If we cut it into three shelves, that would make each shelf _____ feet long.” ____________

5. “The stove is 31 inches wide,” said Annie. “We’ve left a space one yard wide. If we put the stove exactly in the middle, it will leave _____ inches on each side.” ____________

Converting Metric Measurements

The metric system is used all over the world. The three most common units of measure are the gram, the liter, and the meter.

Gram (g) is a measure of weight.
Liter (L) is a measure of volume.
Meter (m) is a measure of length.

The above units are the base units for all other units. The following prefixes can be added to the base units to indicate larger or smaller units:

Kilo means 1,000 base units. It is used to indicate larger measures.

1 kilogram = 1,000 grams
1 kiloliter = 1,000 liters
1 kilometer = 1,000 meters

Centi means 0.01 base unit. It is used only with measures of length.

1 meter = 100 centimeters

Milli means 0.001 base unit. It is used to indicate very small measures.

1 gram = 1,000 milligrams
1 liter = 1,000 milliliters
1 meter = 1,000 millimeters

Examples

Write the appropriate unit of measure for each item.

A. the weight of a car

kilogram

B. the length of a book

centimeter

C. the volume of a spoonful of water

milliliter

In the standard system, a measure may include two units, such as 2 feet 3 inches. In the metric system, the measure is usually written in terms of the larger unit. So, 2 meters 200 millimeters would be written as 2.2 meters.

Use ratios and proportions to convert metric measures.

Examples

D. Convert 47 kilograms to grams.

$$\frac{\text{large unit}}{\text{small unit}} \quad \frac{1\text{ kg}}{1{,}000\text{ g}} = \frac{47\text{ kg}}{n}$$

$$n = \frac{47 \times 1{,}000}{1}$$

$$n = 47{,}000$$

47 kg = 47,000 g

E. Convert 500 centimeters to meters.

$$\frac{\text{large unit}}{\text{small unit}} \quad \frac{1\text{ m}}{100\text{ cm}} = \frac{n}{500\text{ cm}}$$

$$n = \frac{500 \times 1}{100}$$

$$n = 5$$

500 cm = 5 m

Metrics are written in decimal form. Convert all answers to decimals.

F. Convert 4.5 liters to milliliters.

$$\frac{\text{large unit}}{\text{small unit}} \quad \frac{1\text{L}}{1{,}000\text{ mL}} = \frac{4.5\text{L}}{n}$$

$$n = \frac{4.5 \times 1{,}000}{1}$$

$$n = 4{,}500$$

4.5L = 4,500 mL

G. Convert 250 millimeters to meters.

$$\frac{\text{large unit}}{\text{small unit}} \quad \frac{1\text{ m}}{1{,}000\text{ mm}} = \frac{n}{250\text{ mm}}$$

$$n = \frac{1 \times 250}{1{,}000}$$

$$n = 0.25$$

250 mm = 0.25 m

Practice

Write the appropriate metric unit of measure for each item below.

1. the amount of material to make one dress ______________

2. the amount of liquid in an eye dropper ______________

3. the weight of one small hamburger ______________

4. the width of a hair ______________

5. the dimensions of a sheet of paper ______________

6. the amount of gasoline to put in a car ______________

7. the weight of a truckload of gravel ______________

8. the distance between two towns ______________

Convert the following measurements.

9. 37 km = ______ m

10. 4,200 g = ______ kg

11. 2 L = ______ mL

12. 15,000 mm = ______ m

13. 15 cm = ______ m

14. 3.4 g = ______ mg

15. 2.8 m = ______ mm

16. 250 mL = ______ L

Problem Solving

Solve the following problems.

17. Ruth baked a pie weighing 2.4 kilograms. How many grams does it weigh? ______________

18. Alex bought a 2-liter bottle of orange soda. How many milliliters of soda did he buy? ______________

19. If a bag of apples weighs 400 grams, how many kilograms does it weigh? ______________

20. Troy drove his truck 5.8 kilometers to the store, 2.3 kilometers to the gas station, and 1.5 kilometers back home. How many meters did he drive in all?

____

LIFE SKILL

Figuring Meal Portions

"I really like working here in the hospital kitchen," said Frances. "The new responsibility of planning all the meals is exciting. Using metrics to determine portions makes the job easier."

Solve the following problems.

1. "I have 10 liters of orange juice for 40 trays. So each person will get ______ milliliters." ________________

2. "I've made 44 liters of soup. That will serve 160 people, with each person getting ______ milliliters per bowl." ________________

3. "Forty patients ordered chopped steak. Each steak weighs 125 grams. I'll need ______ kilograms of cooked, chopped steak for dinner." ________________

4. "Sixty-five people asked for veal steak. Each piece weighs 72 grams. Each plate gets 2 pieces, so I'll need _____ kilograms of veal." _______________

5. "I have 5 kilograms of cooked liver. Since each serving must be 200 grams, I can serve _____ patients." _______________

6. "Tonight, we have mixed vegetables with 8.5 kilograms of carrots, 5.6 kilograms of peas, 6 kilograms of wax beans, and 7.2 kilograms of lima beans. That will make a total of _____ kilograms of vegetables." _______________

7. "I have 145 orders for potatoes at 200 grams a serving. I'll need _____ kilograms of potatoes tonight." _______________

8. "If each person gets 175 grams of vegetables, I can serve _____ people." _______________

LESSON 23

Working With Metric Measurements

In order to work with metric measurements, change two-unit measurements into decimal values of the larger unit.

Examples

A. Add.
4.7 km + 578 m = ?

$$\frac{1 \text{ km}}{1{,}000 \text{ m}} = \frac{n}{578}$$

Convert 578 meters to kilometers. Set up a proportion and solve for *n*.

$$n = \frac{578 \times 1}{1{,}000}$$

$$n = 0.578 \text{ km}$$

$$\begin{array}{r} 4.7\text{ km} \\ +\ .578\text{ km} \\ \hline 5.278\text{ km} \end{array}$$

Add.

B. Divide.
12.044 kg ÷ 4 = ?

$$\begin{array}{r} 3.011\text{ kg} \\ 4\overline{)12.044\text{ kg}} \end{array}$$

When measures are given in the same unit, solve as you would any decimal problem.

Practice

Convert the smaller unit to a decimal of the larger unit. Then solve.

1. 7.6 kg + 450 g ____________

2. 1.2 m + 800 mm ____________

3. 10.2 L + 800 mL ____________

4. 1.95 m + 5 cm ____________

5. 16 kg − 500 g ____________

6. 90 km − 500 m ____________

7. 800 L − 300 mL ____________

8. 1 m − 1 mm ____________

Solve the following.

9. $\begin{array}{r} 5.045 \text{ kg} \\ \times \quad 8 \\ \hline \end{array}$

10. $5\overline{)4.25 \text{ kg}}$

11. $6\overline{)24.3 \text{ L}}$

12. $\begin{array}{r} 1.01 \text{ g} \\ \times \quad 15 \\ \hline \end{array}$

Problem Solving

Solve the following problems.

13. To make a special punch for 30 people, Salvador needs 4.25 L of ginger ale and 5.8 L of fruit drink. How many liters of punch will that make? ____________

14. Trim for one side of the shed takes 3.3 m. How much trim will be needed for all 4 sides? ____________

15. Divide 6.5 kilograms of apples among 8 containers. What is the total weight of each container of apples if any empty container weighs 500 g? ____________

16. Mary made 1 liter of photographic developer. She used 250 mL to develop 3 rolls of film. How much photographic developer was left over? ____________

17. Twenty-two blood donors are to be served 5.5 L of juice when they are finished. What is the amount of each serving? ____________

LIFE SKILL

Planning a Camping Trip

The Carvers are planning a five-day camping trip in the mountains. It is important for them to know distances, weights, and amounts of supplies. Their survival could depend on having the correct measurements.

Solve the following problems.

1. Each of the 3 adults will carry a backpack that weighs 11 kilograms. Each of the 4 children will carry a 3.5-kilogram pack. What is the total weight of their packs? ________

2. On the first day the group will hike 14.5 kilometers. The second day they will hike 17 kilometers, and the third day they will hike 19.5 kilometers. On the average, how far will they hike each of the first 3 days? ________

3. For dinner one night they will make stew from a package of dried stew. To make 2 servings, they will need to add 450 milliliters of water. How many liters of water will they need to make 10 servings? ________

4. They plan on using most of the supplies by the end of their trip. If each of the 3 adults has a backpack that weighs an average of 0.85 kilograms, and each of the 4 children has a backpack weighing 0.5 kg, what will be the total weight of the backpacks? ________

Problem Solving—Measurements

To solve word problems involving measurements, keep in mind the following steps.

Step 1 Read the problem and underline the key words. These words will generally relate to some mathematics reasoning computation.

Step 2 Make a plan to solve the problem. Ask yourself, Should I add, subtract, multiply, divide, round, or compare? You may have to do more than one of these operations for the same problem.

Step 3 Find the solution. Use your math knowledge to find your answer.

Step 4 Check the answer. Ask yourself, Is this answer reasonable? Did you find what you were asked for?

Be sure your answer is written in the correct units. You may need to convert the units of measure in your answer to the units that were asked for in the problem.

Example

The landscape company uses 40 pounds of decorative gravel per foot of sidewalk. If they used 1,200 pounds of gravel, how many yards long is the sidewalk?

Step 1 The key words are **per foot** and **yards long.** You are to determine the length of the sidewalk in yards.

Step 2 The key words indicate that you should use a proportion to solve this problem. Before you can determine the number of yards in the length of the sidewalk, cross-multiply to determine the number of feet.

$$\frac{40 \text{ lb}}{1 \text{ ft}} = \frac{1{,}200 \text{ lb}}{n}$$

$$\frac{1 \times \overset{30}{\cancel{1{,}200}}}{\underset{1}{\cancel{40}}} = n$$

$$n = 30 \text{ ft}$$

Step 3 Find the solution. The sidewalk is 30 ft long. Convert this measurement to yards using a proportion.

$$\frac{1 \text{ yd}}{3 \text{ ft}} = \frac{n}{30}$$

$$n = \frac{30 \times 1}{3} = 10 \text{ yards}$$

The length of the sidewalk is 10 yards.

Step 4 Check the answer. Does it make sense that 1,200 pounds of gravel could make a sidewalk 10 yards long? Yes, the answer is reasonable.

Practice

Solve the following, using the steps to solve word problems.

1. The disk on the machine spins 30 times a second. How many times does the disk spin in an hour? ____________

2. JJ's Market had 2.25 kg of grapes. The store bought 7.2 kg more grapes, and sold 5.85 kg. How many kilograms of grapes remained at the end of the day? ____________

3. The dress factory uses 5 inches of trim per dress. If 72 dresses are made in one day, how many yards of trim will be used? ____________

4. The restaurant uses $1\frac{1}{2}$ quarts of olive oil a night. If the restaurant is open 6 nights a week, how many gallons will be used in a week? ____________

5. Andre used 2.3 meters of the 5-meter roll of aluminum foil for his art project. How many centimeters of foil were left on the roll? ____________

6. In one day, the factory produces 20,000 feet of fencing. To the nearest whole mile, how many miles of fencing does the factory produce in 5 days? ____________

7. If Ron has 7 guests and 2.1 liters of hot chocolate, how many milliliters can he give to each guest? ____________

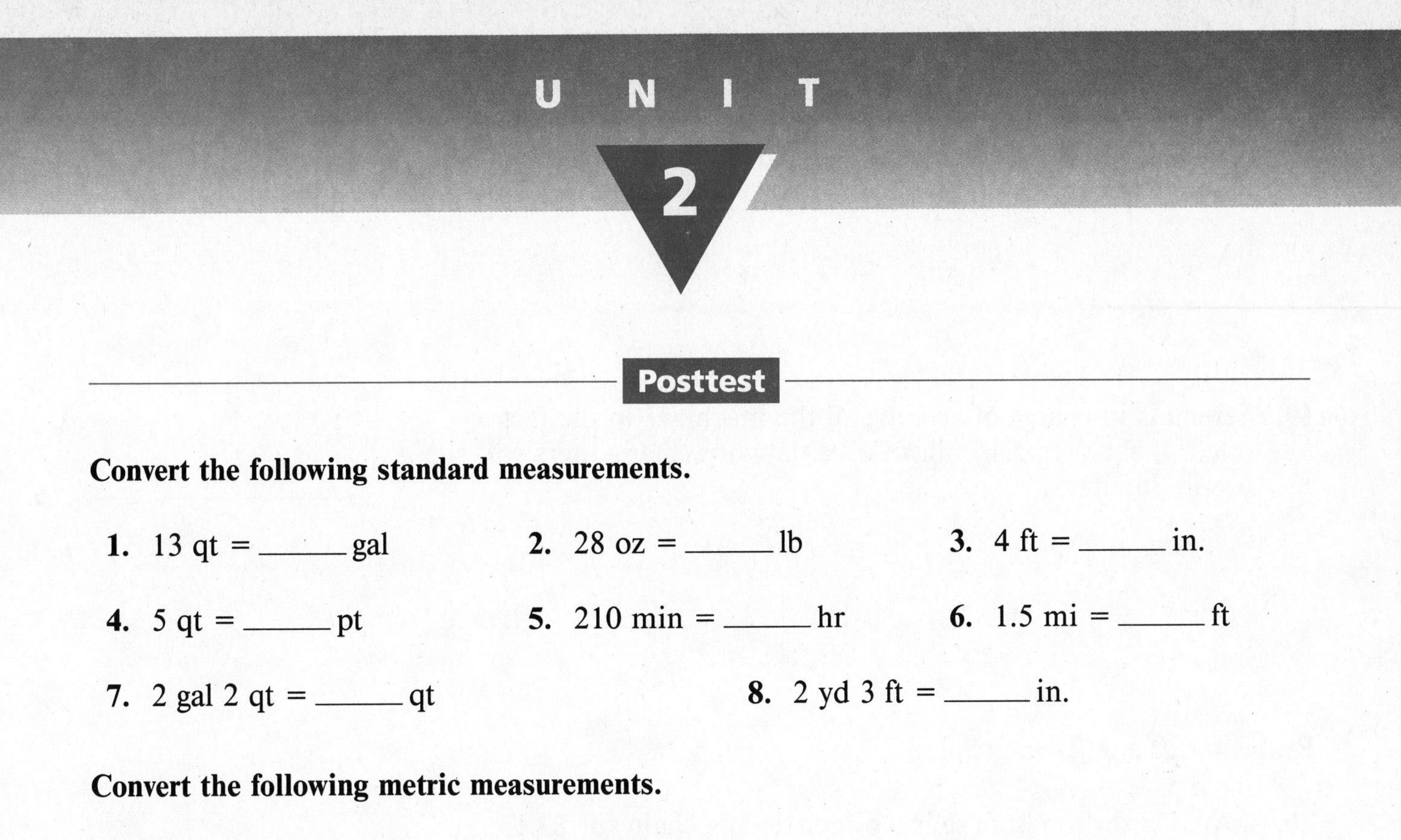

Posttest

Convert the following standard measurements.

1. 13 qt = ______ gal **2.** 28 oz = ______ lb **3.** 4 ft = ______ in.

4. 5 qt = ______ pt **5.** 210 min = ______ hr **6.** 1.5 mi = ______ ft

7. 2 gal 2 qt = ______ qt **8.** 2 yd 3 ft = ______ in.

Convert the following metric measurements.

9. 90 cm = ______ m **10.** 3,300 mL = ______ L **11.** 380 cm = ______ m

12. 71.2 kg = ______ g **13.** 7,000 g = ______ kg **14.** 3.5 L = ______ mL

Problem Solving

Solve the following problems.

15. Mike makes deliveries by truck. First, he drives 4 km 587 m, then he drives 12 km 754 m, and finally 8 km 784 m. How many kilometers does he drive? ______

16. The Morningside Bakery had 16 pounds of flour for bread. After making 10 loaves, they had 4 lb 5 oz of flour left. How much flour did they use? ______

17. The boss wanted Felix to finish 5 projects that were similar in 2 hours 30 minutes. How many minutes should Felix spend on each project? ______

18. Bill bought boards that were 4 feet 2 inches in length to make a bookcase. When he got home, he realized the boards were 8 inches too long. How long should the boards have been? ____________

19. Serena is in charge of keeping all the machines in the factory oiled. If she uses 5.15 mL of oil a day, how many liters will she use in 365 days? ____________

Problems 20 and 21 are related.

20. AAA Hardware Store sells a 6-foot towing chain for $3.15. BBB Home Improvement Center sells towing chain for $0.45 per foot. Which store has the better buy? ____________

21. At AAA Hardware Store, charcoal is $3.95 for a 5-kg bag, and 1 quart of charcoal lighter is $0.95. BBB Home Improvement Center sells charcoal for $0.80 per kilogram, and charcoal lighter for $0.45 a pint. Which store has the better buy on charcoal? ____________

Which store has the better buy on charcoal lighter? ____________

UNIT 3

Picture Graphs and Tables

Use the picture graph below to answer questions 1–5.

Guests at Hotels During the Month of July	
Parker Hotel	
Robertson Hotel	
Smith Hotel	
Beck Hotel	

= 250 Guests

1. How many people stayed at the Parker Hotel? ________

2. How many more people stayed at the Beck Hotel than at the Smith Hotel? ________

3. The number of guests who stayed at the Smith Hotel are what percent of the number of people who stayed at the Parker Hotel? ________

4. What was the total number of guests at the four hotels? ________

5. What was the average number of guests per hotel? ________

Use the figure below to answer the following questions.

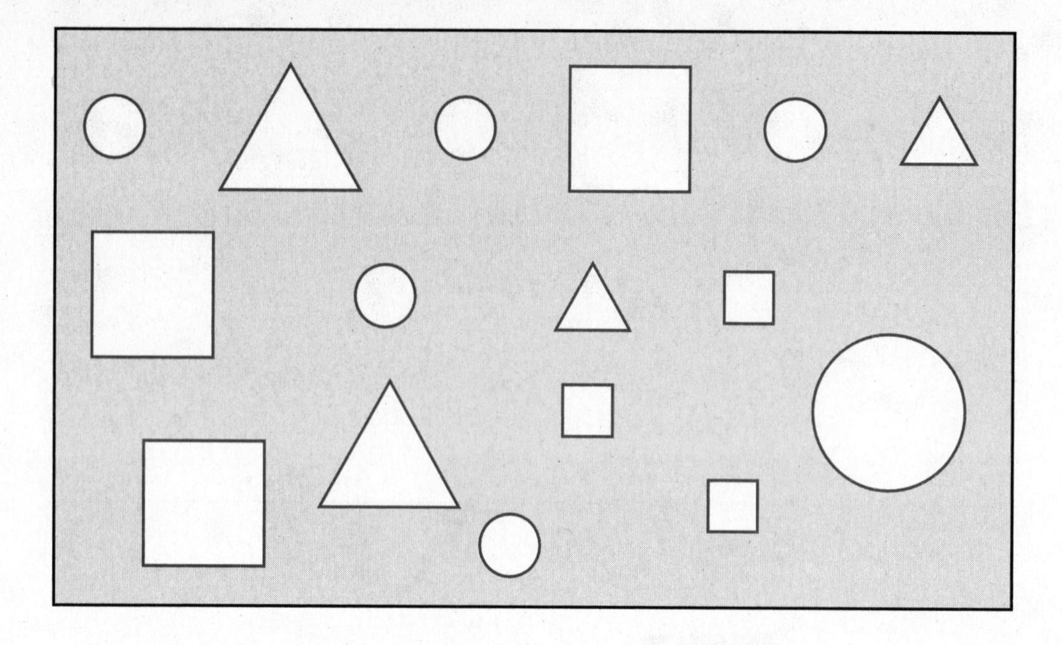

6. Count and categorize the shapes. Then complete the table.

Shapes

	Circles	Squares	Triangles	Total
Large				
Small				
Total				

7. How many more small circles are there than large circles? ____________

8. What is the ratio of small squares to small triangles? ____________

9. The large circles are what fraction of all the figures? ____________

10. The triangles are what percent of all the figures? ____________

11. The large figures are what fraction of the small figures? ____________

LESSON 25

Picture Graphs

Picture graphs compare information by using symbols. A **symbol** is a picture that stands for something. The **key** tells you what the symbols in the graph stand for. It is usually near the graph. When reading a picture graph, follow these steps:

Step 1 Read the title of the graph.

Step 2 Find what things are being compared.

Step 3 Look at the key to see what the symbols stand for.

> **MATH HINT**
>
> If part of a symbol appears in a picture graph, estimate the amount or the number of things that it stands for.

Example

Use the steps above to read the following picture graph.

J. R. Lowden Company Sales

Pepé	\$ \$ \$ \$
José	\$ \$ \$
Juanita	\$ \$ \$ \$ \$
Carlotta	\$ \$ \$

\$ = \$5,000 in sales

Step 1 Read the title.
The title of this graph is **J.R. Lowden Company Sales.**

Step 2 Find what is being compared.
The amount of sales of four people in the company is being compared.

Step 3 Look at the key to see what the symbol stands for.
The key indicates that one symbol stands for \$5,000 in sales. One-half a symbol stands for one-half of \$5,000, or \$2,500.

Practice

Use the following picture graph to answer questions 1–6.

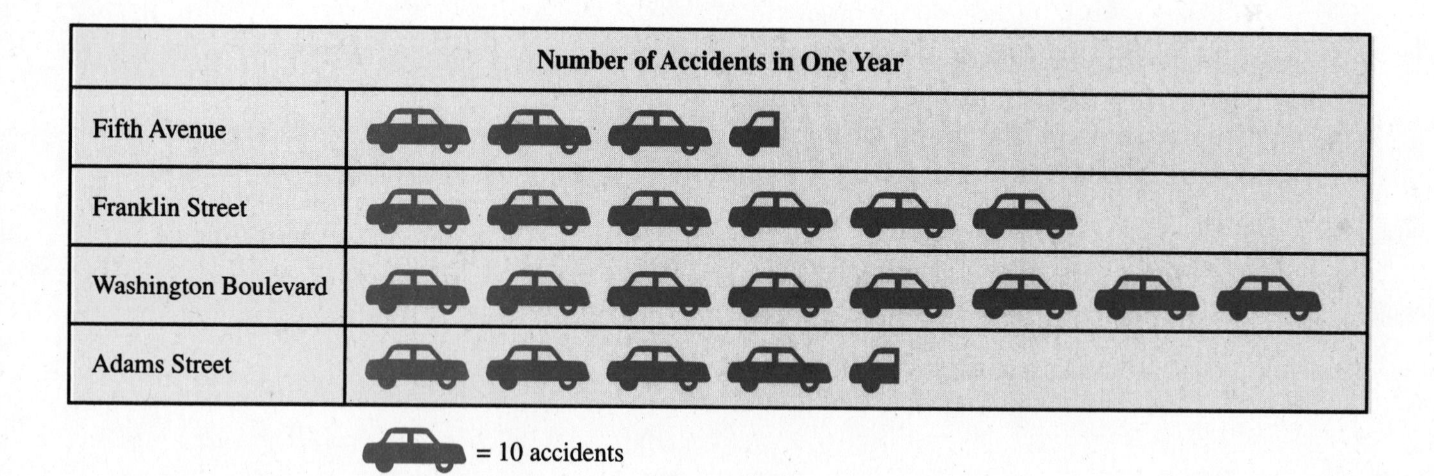

1. What is the title of this graph?

____

2. What does each symbol represent?

____

3. What is being compared in this graph?

____

4. How many accidents occurred on Washington Boulevard in one year?

____

5. How many accidents occurred on Adams Street in one year?

____

6. If Union Street had 30 accidents, how many symbols would be used?

____

Use the following picture graph to answer questions 7–10.

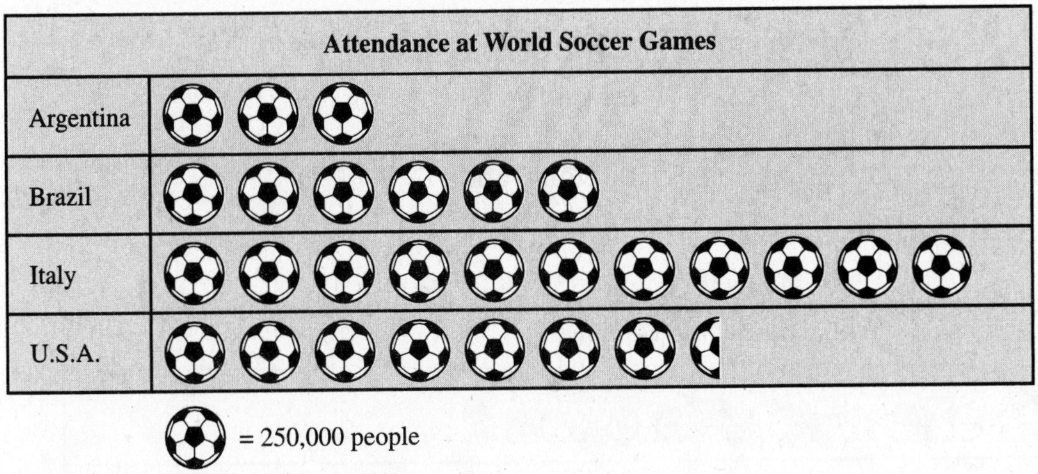

Attendance at World Soccer Games
Argentina
Brazil
Italy
U.S.A.

= 250,000 people

7. What does each symbol represent?

____

8. What is being compared in this graph?

____

9. How many people attended soccer games in Brazil?

____

10. If Germany had 1,000,000 people attending their games, how many symbols would be used?

____

LESSON 26

Analyzing Picture Graphs

You can find information by analyzing picture graphs. You can also compare information using subtraction, ratios, fractions, and percents.

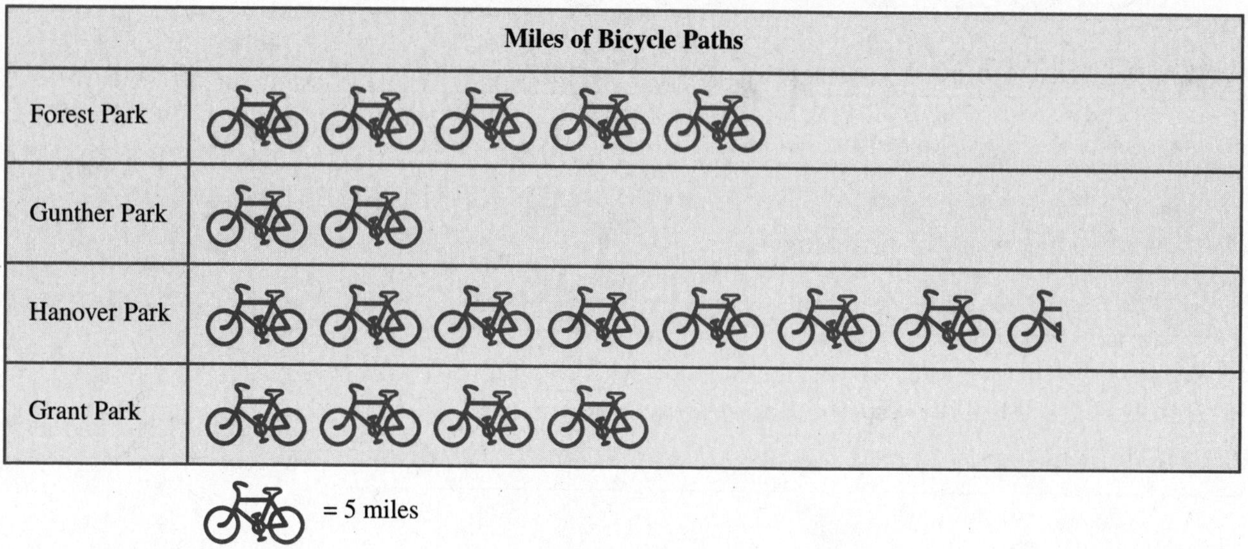

Miles of Bicycle Paths	
Forest Park	
Gunther Park	
Hanover Park	
Grant Park	

= 5 miles

Examples

Use the picture graph above to solve the problems.

A. How many more miles of bicycle paths are in Hanover Park than in Gunther Park?

Hanover Park has $7\frac{1}{2}$ symbols.
To find the number of miles, multiply.
$7\frac{1}{2} \times 5 = \frac{15}{2} \times 5 = \frac{75}{2} = 37\frac{1}{2}$ miles

Gunther Park has 2 symbols.

$2 \times 5 = 10$ miles
$37\frac{1}{2} - 10 = 27\frac{1}{2}$ miles   Subtract.

Hanover Park has $27\frac{1}{2}$ more miles of bicycle paths than Gunther Park.

B. The miles of paths in Grant Park are what fraction of the miles of paths in Forest Park?

Grant Park has 4 symbols.
To find the number of miles, multiply.
$4 \times 5 = 20$ miles

Forest Park has 5 symbols.

$5 \times 5 = 25$
$\frac{20}{25} = \frac{4}{5}$ Write as a fraction in lowest terms.

The miles of paths in Grant Park are $\frac{4}{5}$ of the miles of paths in Forest Park.

Practice

Use the following picture graph to answer questions 1–6.

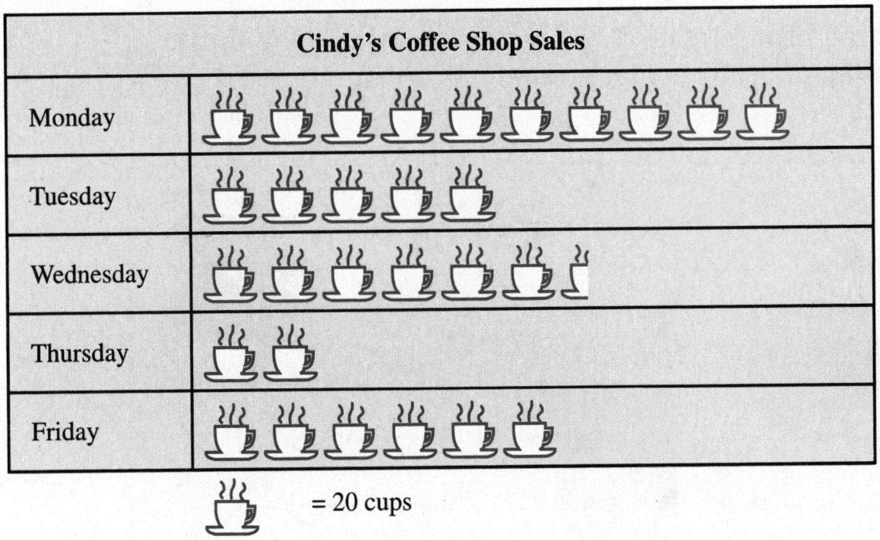

Cindy's Coffee Shop Sales	
Monday	
Tuesday	
Wednesday	
Thursday	
Friday	

= 20 cups

1. How many cups of coffee were sold on Monday?

   ____

2. How many more cups of coffee were sold on Wednesday than on Friday?

   ____

3. The number of cups of coffee sold on Thursday is what fraction of the number sold on Monday?

   ____

4. The number of cups sold on Monday is how many times the number sold on Tuesday?

   ____

Problem Solving

Solve the following problems.

5. Cindy wants to order coffee for next week. She expects her coffee sales to be approximately the same as this week. If one pound of coffee makes 50 cups, how many pounds of coffee should she order?

   ____

6. If Cindy charges $0.89 per cup of coffee, how much money would she make in 8 weeks, based on the sales in the graph?

   ____

Organizing Data

Data is a collection of information. To understand data, you must organize it. The figure below contains data.

To organize data, follow these steps:

Step 1 Title the data.

Step 2 Choose the categories.

Step 3 Group the data into categories by counting the number in each.

Examples

A. Organize the data in the figure above.

Step 1 Title this group of data "Shapes."

Step 2 Choose circles, squares, and triangles as your categories.

Step 3 There are 2 circles, 4 squares, and 6 triangles.

B. Look at the figure below. It contains categories of shapes, including circles, squares, and triangles. It also contains categories of large and small figures.

Step 1 Title the data. Call this set of data "Shapes."

Step 2 Choose the categories. There are two categories. The first contains shapes: circles, squares, and triangles. The second contains sizes: large and small.

Step 3 Group the data into categories by counting the number in each. To make this step easier, use the box of figures to answer the following questions.

How many squares are	large? _____	small? _____	altogether? _____
How many circles are	large? _____	small? _____	altogether? _____
How many triangles are	large? _____	small? _____	altogether? _____
How many figures are	large? _____	small? _____	altogether? _____

Use the figure below to answer questions 1–3.

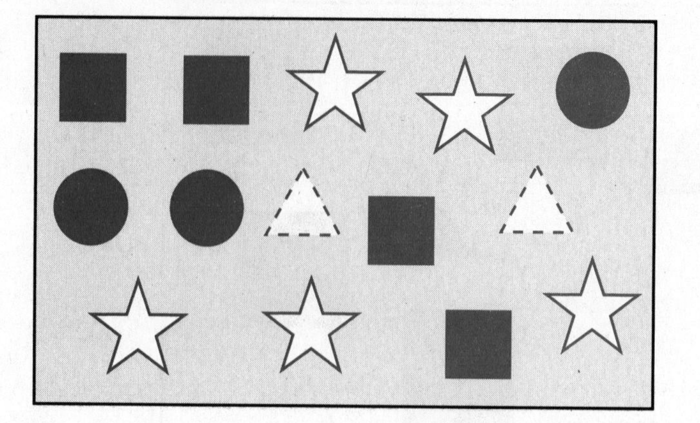

1. Give the data a title.

2. What are the categories?

3. How many figures are in each category?

Use the figure below to answer questions 4 and 5.

4. Give the data a title.

5. One category is shapes. What is the other?

Use the figure below to answer questions 6–8.

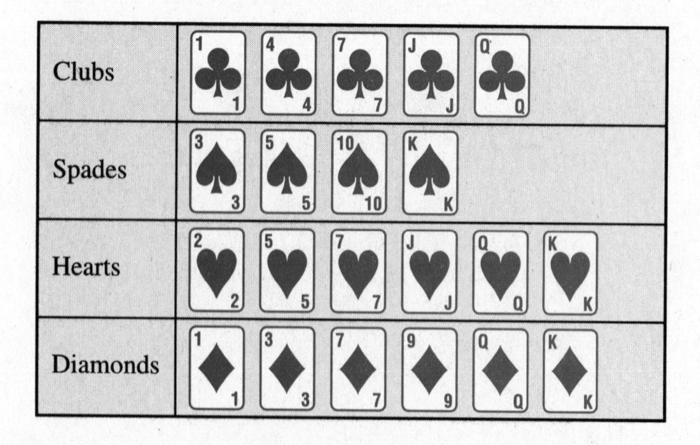

6. Give the data a title.

7. What are the categories?

8. How many figures are in each category?

Organizing Data Into Tables

Once you organize data, you can put the information into a table. A **table** is one way to show two categories at the same time. The **rows** make up one category. The **columns** make up the other. When a row and a column intersect, they create a box, or cell. The information in that box, or cell, is in both categories. To organize data into tables, follow these steps:

Step 1 Give the table a title.

Step 2 Write the names of one category in the first column. Write the names of the other category in the top row.

Step 3 Count the number of data in each category and write the information in the table.

The names of the categories in the top row of a table are called **headings.**

Example

Organize the following data into a table.

Step 1 Give the table a title—"Shapes."

Step 2 Write the names of one category in the first column. Write the names of the other category in the first row.

Step 3 Count the number in each category. Fill in the table.

Shapes

	Triangle	**Square**	**Circle**	**Total**
Large	5	1	3	9
Small	5	4	2	11
Total	10	5	5	20

Practice

Make a table for each of the following sets of data.

1.

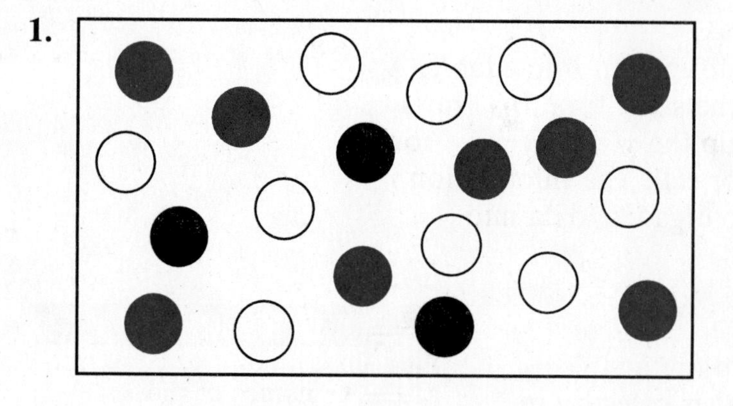

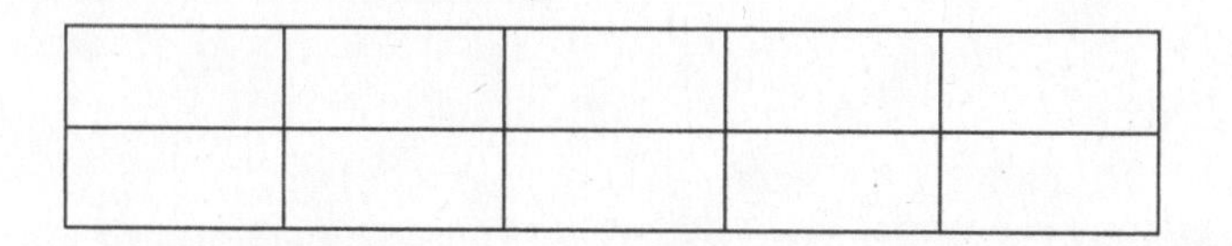

2.

LESSON 29

Analyzing Data in Tables

You can analyze data once it is organized into tables.

Example

The table below lists the prices of theater tickets. The prices are different according to two categories: the age of the buyer and the location of the seats. The Yinglings would like to sit in the 15th row. What is the cost of two adult tickets?

BelAire Theater Ticket Prices for the Musical Review				
	Rows 1–10	Rows 11–24	Rows 25–40	Rows 41–56
Children Under 12	$20	$15	$12	$10
Adults	$40	$30	$24	$20
Senior Citizens 65 and Older	$25	$20	$16	$15

To find out the cost of two adult tickets in Row 15, look for the column marked Rows 11–24 and the row marked Adults.

The intersection of the row and column shows $30. One ticket costs $30. Two tickets would be $\$30 \times 2 = \$60$.

Practice

Use the table above to answer questions 1–5.

1. What is the price of a senior citizen ticket in row 27? ______
2. How much more expensive are four adult tickets in rows 1–10 than in rows 41–56? ______
3. Four men and five women in the Golden Years Club for senior citizens want to attend the show. They want to sit in rows 25 through 40. What will their tickets cost? ______
4. Three high school students want the cheapest tickets they can get. What will their tickets cost? ______
5. What is the cost of tickets for a family of four, two adults and two children, who would like to sit in one of the first ten rows? ______

The table below shows the recipes used for a pancake mix. Use the information in the table to answer questions 6–12.

Recipe	Number of Servings	Pancake Mix	Milk	Eggs	Additional Ingredients
Traditional Pancakes	3	2 cups	1 cup	2	—
Traditional Pancakes	6	4 cups	2 cups	4	—
Thin Pancakes	3	2 cups	$1\frac{1}{2}$	2	2 Tbs. sugar
Waffles	3	2 cups	$1\frac{1}{3}$	1	2 Tbs. oil

6. Are any additional ingredients needed for traditional pancakes? ____________

7. Helouise has only one egg. Using the pancake mix, what should she make for breakfast? ____________

8. How much more milk is needed to serve waffles to 3 people than to serve traditional pancakes to 3 people? ____________

9. How many eggs are needed to make 21 servings of thin pancakes? ____________

10. How much pancake mix is needed to serve 24 people thin pancakes? ____________

11. If a box of mix will make 33 servings of traditional pancakes, how many cups of mix are in the box? ____________

12. Juana is serving breakfast to a total of 10 people. Two are children. Together the children will only eat one serving. How much milk will she need to make waffles?

____

LIFE SKILL

Reading a Nutrition Table

Nutrition tables like the one shown at right appear on containers of food sold in stores. They provide nutrition facts about the food inside the containers.

Nutrition Facts for Crunchy Chips

Serving Size About 7 chips (28 g)
Servings Per Container 14

Amount Per Serving	
Calories 150	Calories from Fat 25
	% Daily Value*
Total Fat 7.8 g	**12**%
Saturated Fat 1.5 g	**7.5**%
Cholesterol 0 mg	**0**%
Sodium 72 mg	**3**%
Total Carbohydrate 18 g	**6**%
Dietary Fiber 1 g	**4**%
Sugars 0 g	
Protein 2 g	

* Percent Daily Values are based on a 2,000 calorie diet. Your daily values may be higher or lower depending on your calorie needs.

Use the nutrition table above to solve the following problems.

1. What is the total number of calories per serving?

____

2. How many milligrams of sodium are in one serving?

____

3. In one serving, how many more grams of fat are there than of protein?

____

4. What is the percent daily value of dietary fiber in one serving?

____

5. If Josie ate two servings of chips, how many grams of carbohydrate would she have eaten?

____

6. If Joe ate four servings of chips, what would be the percent daily value of sodium?

____

LIFE SKILL

The table labeled "Recommended Intake" shows the amount of fat, cholesterol, sodium, and carbohydrates for people whose average daily calorie intake is 2,000 or 2,500 a day. The other three tables show the nutrition facts for chicken breast patties, potatoes, and peas.

Recommended Intake

	Calories:	2,000	2,500
Total Fat	Less than	65 g	80 g
Saturated Fat	Less than	20 g	25 g
Cholesterol	Less than	300 mg	300 mg
Sodium Total	Less than	2,400 mg	2,400 mg
Carbohydrate		300 g	375 g
Dietary Fiber		25 g	30 g

Nutrition Facts for Chicken Breast Patties

Serving Size 1 patty (75 g)

Calories 190

	% Daily Value
Total Fat 19 g	**29%**
Saturated Fat 3 g	**13%**
Cholesterol 30 mg	**11%**
Sodium 230 mg	**10%**
Total Carbohydrate 11 g	**4%**
Dietary Fiber 1 g	**4%**

Nutrition Facts for Potatoes

Serving Size 84 g

Calories 110

	% Daily Value
Total Fat 2.5 g	**4%**
Saturated Fat 1 g	**5%**
Cholesterol 0 mg	**0%**
Sodium 15 mg	**.6%**
Total Carbohydrate 19 g	**6%**
Dietary Fiber 2 g	**8%**

Nutrition Facts for Peas

Serving Size 1/2 cup (89 g)

Calories 70

	% Daily Value
Total Fat 0 g	**0%**
Saturated Fat 0 g	**0%**
Cholesterol 0 mg	**0%**
Sodium 125 mg	**5%**
Total Carbohydrate 13 g	**4%**
Dietary Fiber 5 g	**20%**

Use the four tables on page 84 to answer questions 7–9.

For dinner, Helena is going to have one chicken breast patty, two servings of potatoes, and one serving of peas.

7. What is the percent daily value of sodium of the meal? ______________

8. If Helena normally eats 2,000 calories a day, how many more grams of carbohydrate can she eat? ______________

9. If swimming burns or consumes 6 calories per minute, how long would Helena have to swim to burn all the calories of her dinner? ______________

Problem Solving—Tables

Tables are useful in helping you to find information and solve problems. Remember the following steps to solve word problems:

Step 1 Read the problem and underline the key words. These words will generally relate to some mathematics reasoning computation.

Step 2 Make a plan to solve the problem. Ask yourself, Should I add, subtract, multiply, divide, round, or compare? You may have to do more than one of these operations for the same problem. You may also be able to estimate your answer.

Step 3 Find the solution.

Step 4 Check the answer. Ask yourself, Is this answer reasonable? Did you find what you were asked for?

Example

Maria is going to a job interview. She wants to arrive at the bus stop Reed Rd. at Hendley Rd. before 2:00 p.m.. What is the latest westbound bus she can catch from Morton Rd.? Use the bus schedule below.

MATH HINT

The left table shows buses traveling west. The right table shows buses traveling east.

MONDAY-FRIDAY WEST

FIELD RD. at KARL RD.	KARL RD. at NORTHTOWN	MORTON RD. at HIGH ST.	HENDLEY RD. at RIVER RD.	BETHEL RD. at DOWNY RD.	REED RD. at HENDLEY RD.	KENZIE DR. at REED RD.
◀ 8:10	8:12	8:19	8:23	8:30	8:36	8:37
◀ 9:26	9:29	9:37	9:42	9:49	9:55	9:56
◀10:46	10:49	10:57	11:02	11:09	11:15	11:16
◀12:06	**12:09**	**12:17**	**12:22**	**12:29**	**12:35**	**12:36**
◀ 1:26	**1:29**	**1:37**	**1:42**	**1:49**	**1:55**	**1:56**
◀ 2:46	**2:49**	**2:57**	**3:02**	**3:09**	**3:15**	**3:16**
◀ 4:06	**4:09**	**4:17**	**4:22**	**4:29**	**4:35**	**4:36**
◀ 5:26	**5:29**	**5:37**	**5:42**	**5:49**	**5:55**	**5:56**

◀Wheelchair lift-equipped trip

MONDAY-FRIDAY EAST

KENZIE DR. at REED RD.	HENDLEY RD. at REED RD.	DOWNY RD. at BETHEL RD.	HENDLEY RD. at RIVER RD.	HIGH ST. at MORTON RD.	NORTHTOWN SHOPPING CENTER	FIELD RD. at KARL RD.
◀ 8:43	8:47	8:52	9:00	9:06		9:14
◀10:03	10:07	10:13	10:23	10:29	10:37	10:40
◀11:23	11:27	11:33	11:43	11:49	11:57	**12:00**
◀12:43	**12:47**	**12:53**	**1:03**	**1:09**	**1:18**	**1:21**
◀ 2:03	**2:07**	**2:13**	**2:23**	**2:29**	**2:38**	**2:41**
◀ 3:23	**3:27**	**3:33**	**3:43**	**3:49**	**3:58**	**4:01**
◀ 4:43	**4:47**	**4:53**	**5:03**	**5:09**	**5:18**	**5:21**
◀ 6:03	**6:07**	**6:13**	**6:23**	**6:29**	**6:37**	

Light type a.m. **Bold type p.m.**

Step 1 Maria must determine the latest bus she can catch from Morton Rd. to arrive at Reed Rd. before 2:00 p.m. The key words are **latest** and **before 2:00 p.m.**

Step 2 The key words indicate what you should look for when reading the tables.

Step 3 Find the solution. Look in the column of the first table labeled "Reed Rd. at Hendley Rd." The latest arrival time before 2:00 is 1:55. Now move left in that same row until you reach the column labeled "Morton Rd. at High St." The time is 1:37. Therefore, the latest bus Maria can catch is at 1:37 p.m.

Practice

Solve the following problems using the bus schedule tables on page 86 and the steps for solving word problems.

1. What is the earliest eastbound bus leaving the Kenzie Dr. at Reed Rd. bus stop? __________

2. If Jose catches the 2:07 p.m. eastbound bus at the Hendley Rd. at Reed Rd. bus stop, what time will he arrive at High St.? __________

3. Terry gets out of work at 4:00 p.m. If he walks 12 minutes to the bus stop at Karl Rd. at Northtown, what is the earliest bus he can catch traveling west? __________

4. Natasha has a doctor appointment downtown. If she needs to arrive at the High St. at Morton Rd. bus stop before 11:00 a.m. to get to her appointment on time, what is the latest bus she can take from Kenzie Dr.? __________

5. Olivia takes the bus each morning at 8:10 from her home on Field Rd. to her job on Kenzie Dr. She also takes the bus home every evening. If she works five days a week, how many hours does she spend each week traveling on the bus to and from work? __________

6. Kendra is going to take the 10:13 bus from Downy Rd. and travel east to Northtown Shopping Center. She wants to spend 2 hours shopping and then take the bus to visit her friend on Reed Rd. It takes 8 minutes to walk from the bus stop on Reed Rd. to her friend's house. What is the earliest time she will arrive at her friend's house? __________

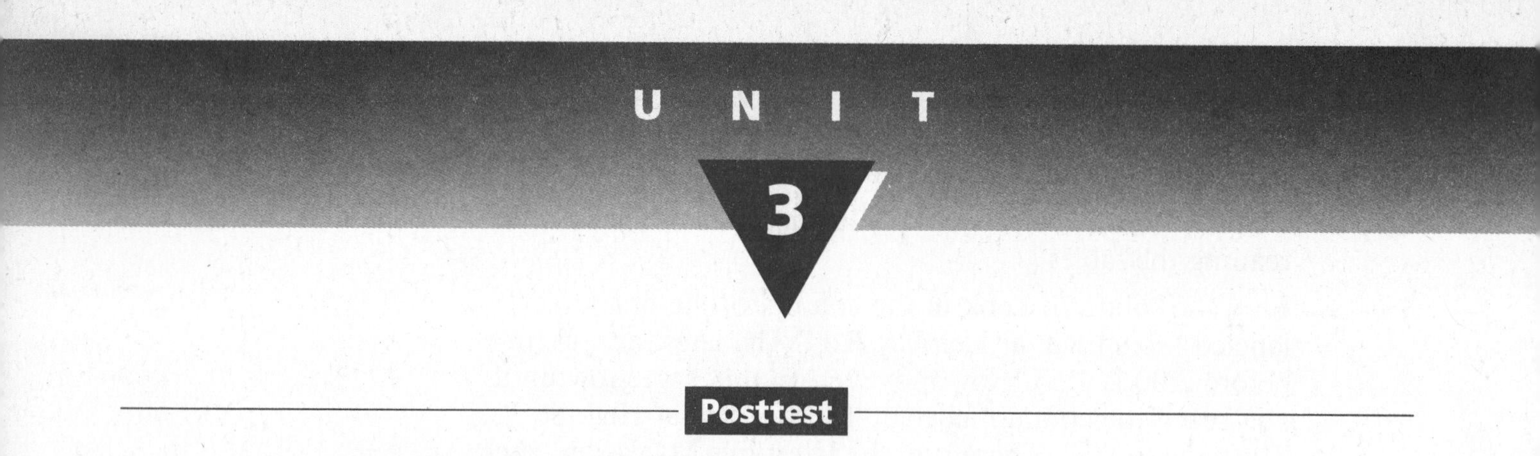

Posttest

Use the picture graph below to answer questions 1–6.

1. How many complaints did Company D receive? __________

2. How many symbols would be used to represent 360 complaints? __________

3. The number of complaints for Company A is what percent of the complaints for Company D? __________

4. The number of complaints for Company B is what fraction of the complaints for Company C? __________

5. Company A is what percent of the total number of complaints for the four companies? __________

6. What is the average number of complaints per company? __________

Use the following table to answer questions 7–14.

Home Service Usage Rate Chart		Price Per Minute		
Miles	Duration Of Call	Mon-Fri 9 a.m.-11 a.m. 2 p.m.-8 p.m.	Mon-Fri 8 a.m.-9 a.m. 11 a.m.-2 p.m. & 8 p.m.-9 p.m.	Mon-Fri 9 p.m.-8 a.m. All Hours Sat-Sun & Specified National Holidays*
Your Local Calling Area	Untimed	5.2¢	4.7¢	3.1¢
8–15	First Minute	8.0¢	7.2¢	4.8¢
	Additional Minutes	2.3¢	2.1¢	1.4¢
16–40	First Minute	10.4¢	9.4¢	6.2¢
	Additional Minutes	3.4¢	3.1¢	2.0¢
40+	First Minute	16.3¢	14.7¢	9.8¢
	Additional Minutes	6.5¢	5.9¢	3.9¢

7. What is the title of the table? __________

8. What are the four categories of distances listed that affect the price of a phone call? __________

____

____

____

9. For a phone call outside your local calling area, what are the two categories for the duration of a call? __________

____

10. What is the cost of a three-minute call 13 miles away on a Tuesday at 10:00 a.m.? __________

11. What is the cost of the same call at 12:15 p.m.? __________

12. What is the cost of an hour-long call made to a town 42 miles away at 11:00 p.m. on a Friday? __________

13. What is the cost of a 25-minute call to a home in the local dialing area made on a Saturday? __________

14. How much more expensive is a 10-minute call at 9:00 a.m. on a Wednesday to a town 16 miles away than a 15-minute call made at 11:30 a.m. on a Saturday to the same town? __________

UNIT

4

Bar Graphs and Line Graphs

Pretest

Refer to the graph to answer questions 1–4.

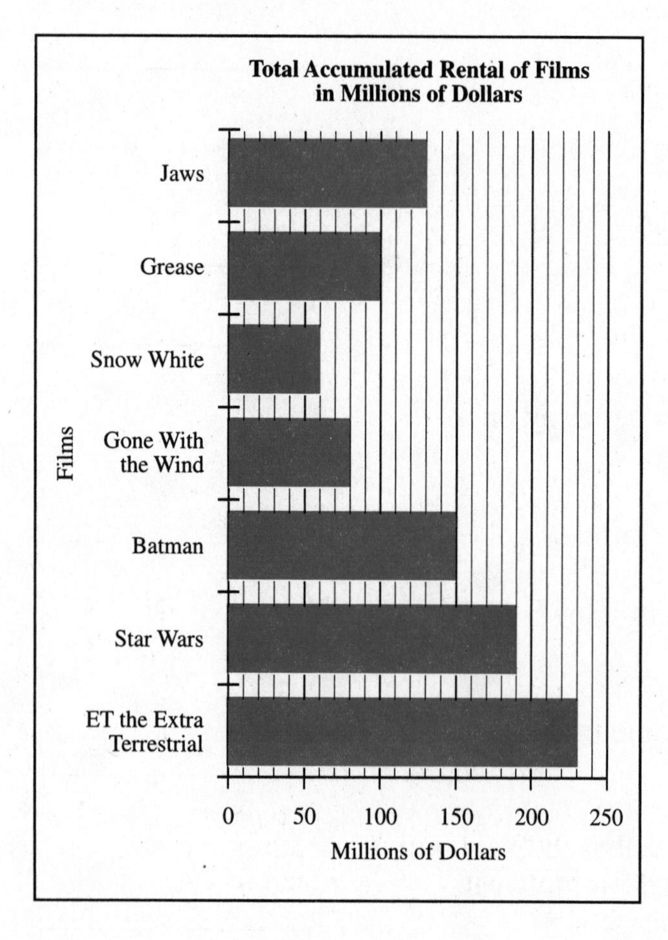

1. How many dollars have been spent renting *ET*?

   ____

2. How many more dollars have been spent renting *Star Wars* than have been spent renting *Grease*?

   ____

3. If *Gone With the Wind* was made 50 years ago, what is the average number of dollars spent per year renting the film?

   ____

4. The amount of dollars earned by *Grease* is what fraction of the money earned by *Batman*?

   ____

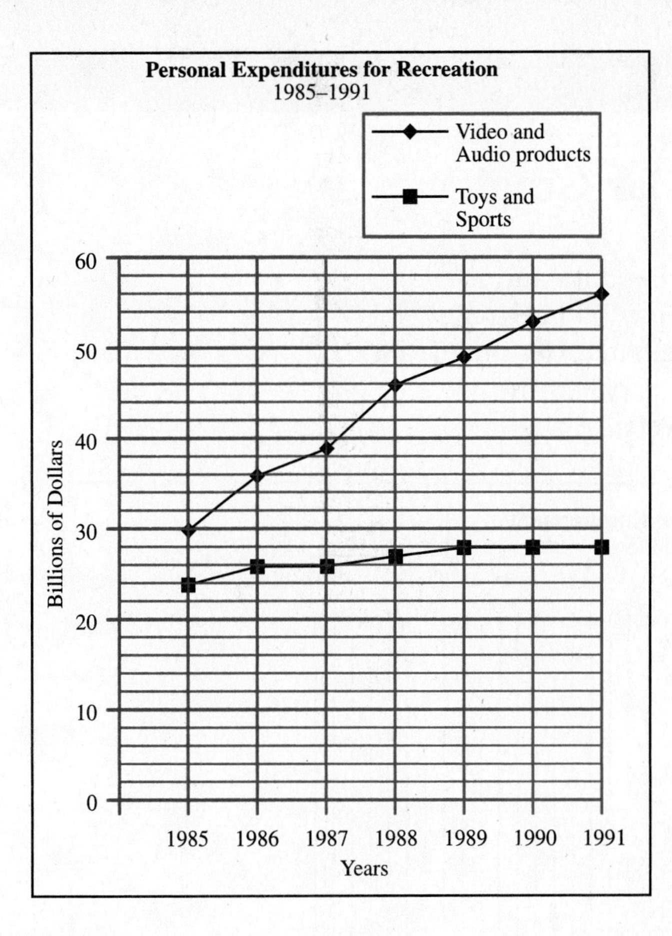

Refer to the graph at the left to answer questions 5–8.

5. What is the difference in personal expenditures between audio and video equipment and toys and sports in 1988?

____

6. The sale of toys and sports products is what percent of the sale of audio and video products in 1985?

____

7. The sale of toys and sports products is what percent of the sale of audio and video products in 1991?

____

8. What is the percent of increase in sales of audio and video products between 1985 and 1987?

____

Refer to the graph at the right to answer questions 9–12.

9. The heights of the waterfalls in Norway is what fraction of the heights of the waterfalls in the U.S.?

____

10. What is the difference in height between the waterfalls in Norway and the ones in Australia?

____

11. The heights of the waterfalls in the U.S. are how many times greater than the heights of the waterfalls in India?

____

12. What is the difference in height between the highest waterfall and the shortest waterfall?

____

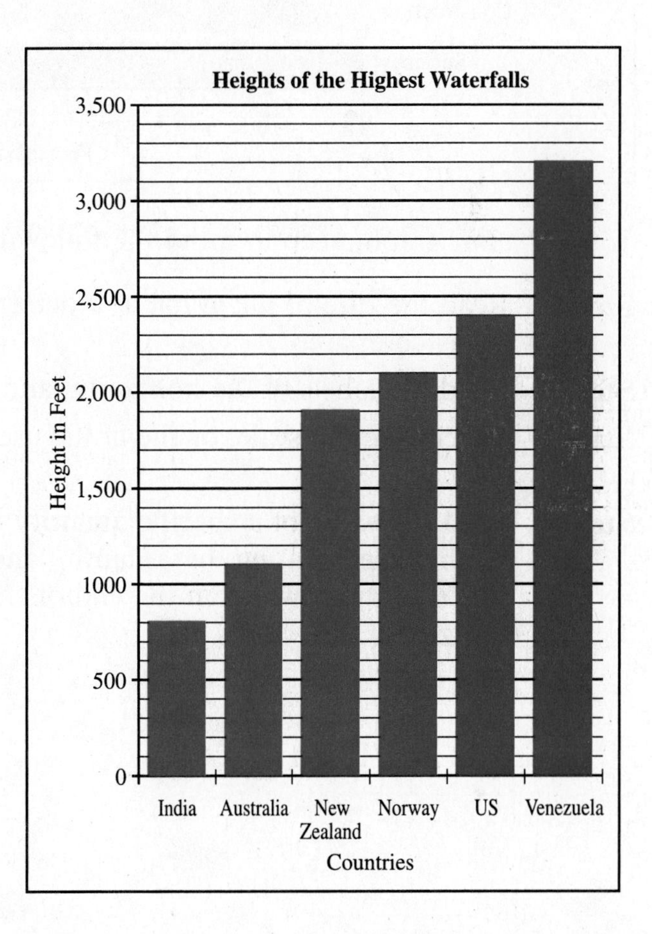

Reading a Bar Graph

A bar graph is useful for organizing and displaying data and for showing how sets of data compare. Bar graphs may be **horizontal,** like the graph shown below, or **vertical,** with the bars extending from the horizontal axis. The horizontal number line is called the horizontal axis; the vertical number line is the vertical axis.

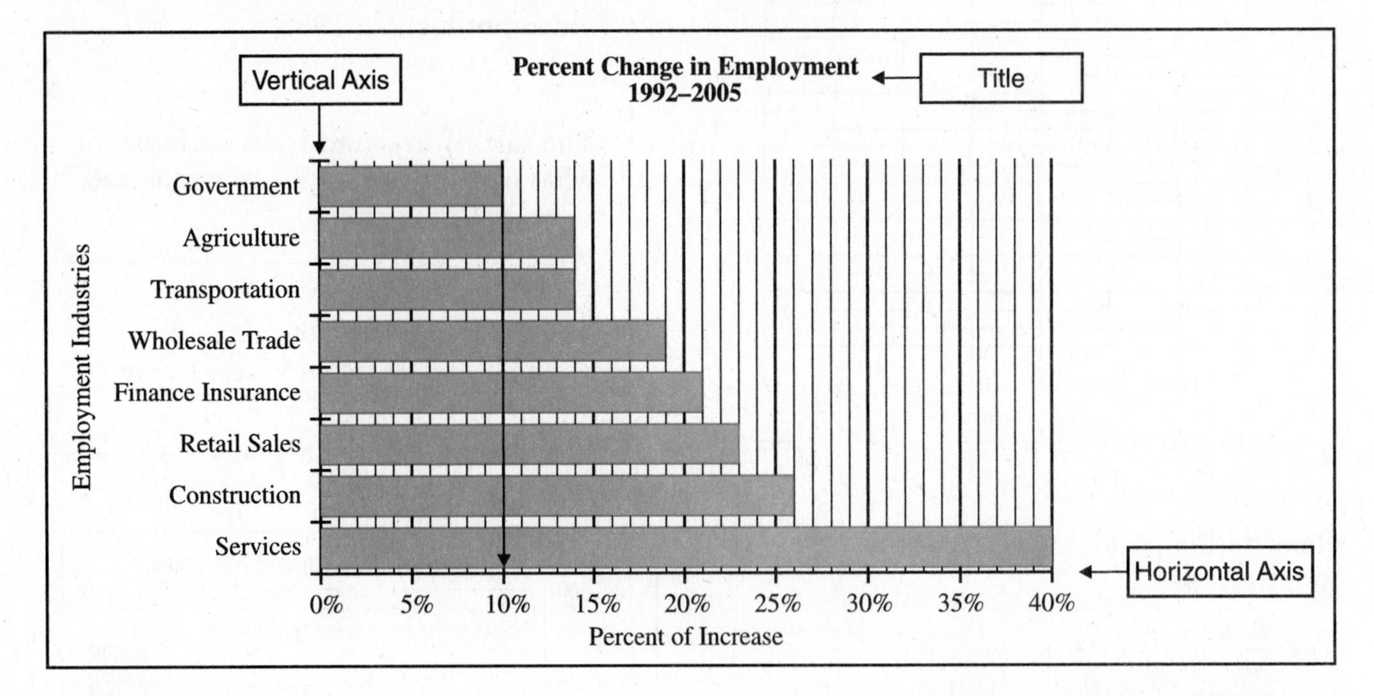

To read a bar graph, keep in mind the following steps:

Step 1 Read the title of the graph and determine what is being compared.

Step 2 Read the labels of the horizontal and vertical axes.

Step 3 Determine the scale, or the units, used along the axes.

Step 4 Find the value of a specific quantity by looking for the bar representing that quantity and drawing a line straight down from the end of the bar to the horizontal axis.

MATH HINT

On a vertical bar graph, you find the value of a specific quantity by drawing a line straight across from the end of the bar to the vertical axis.

Example

Refer to the graph on page 92.

By what percent is employment in government expected to increase between 1992 and 2005?

Step 1	Read the title of the graph and determine what is being compared.	The title is "Percent Change in Employment, 1992–2005." The graph is comparing the percent of increase of various employment industries from 1992 to 2005.
Step 2	Read the labels of the horizontal and vertical axes.	The label of the vertical axis is "Employment Industries" and the label of the horizontal axis is "Percent of Increase."
Step 3	Determine the scale or the units used along the axes.	Each unit on the horizontal axis represents 1% increase in employment.
Step 4	Find the value of a specific quantity by looking for the bar representing that quantity and drawing a line straight down from the end of the bar to the horizontal axis.	Look for the bar labeled "Government." Draw a line straight down from the end of the bar. Read the percentage value. The percentage value is 10%.

Employment in government is expected to increase by 10% between 1992 and 2005.

Practice

Use the bar graph below to answer questions 1–5.

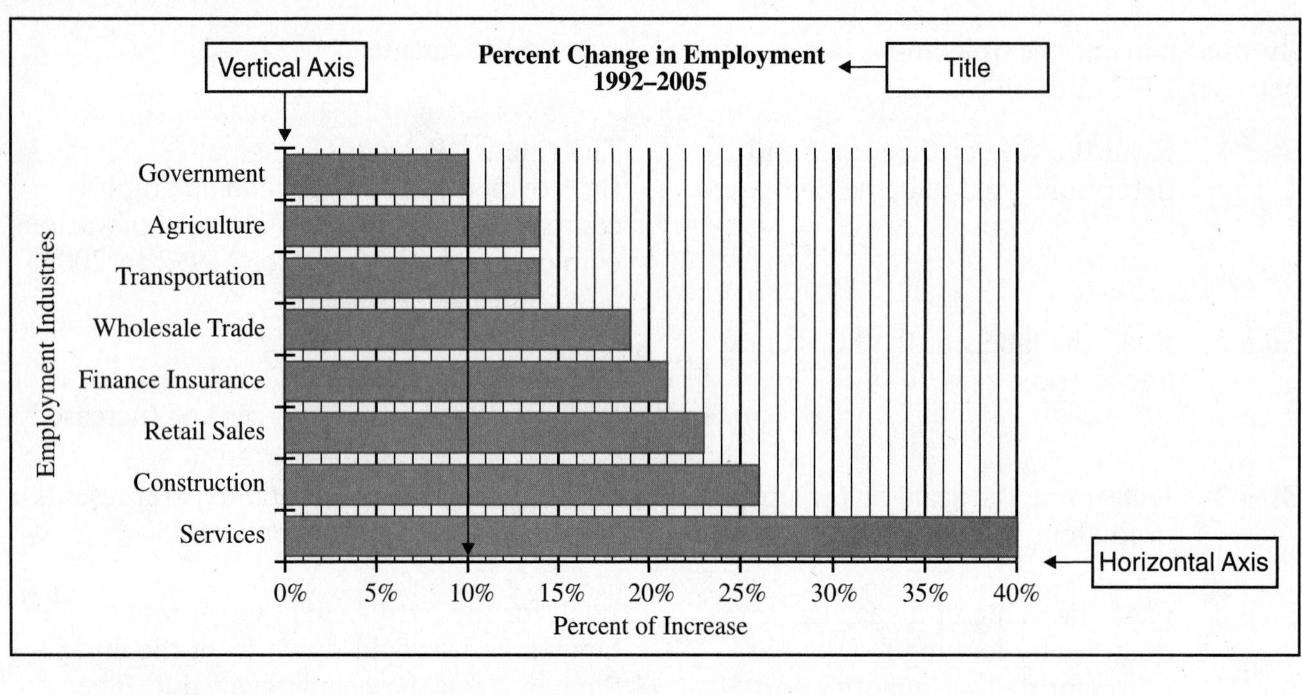

1. The finance insurance industry is expected to increase by what percent by the year 2005? __________

2. The transportation industry is expected to increase by what percent by the year 2005? __________

3. Which two industries are expected to increase at the same rate? __________

4. Which industry is expected to have a 40% increase in employment? __________

5. Which industry is expected to have a 21% increase in employment? __________

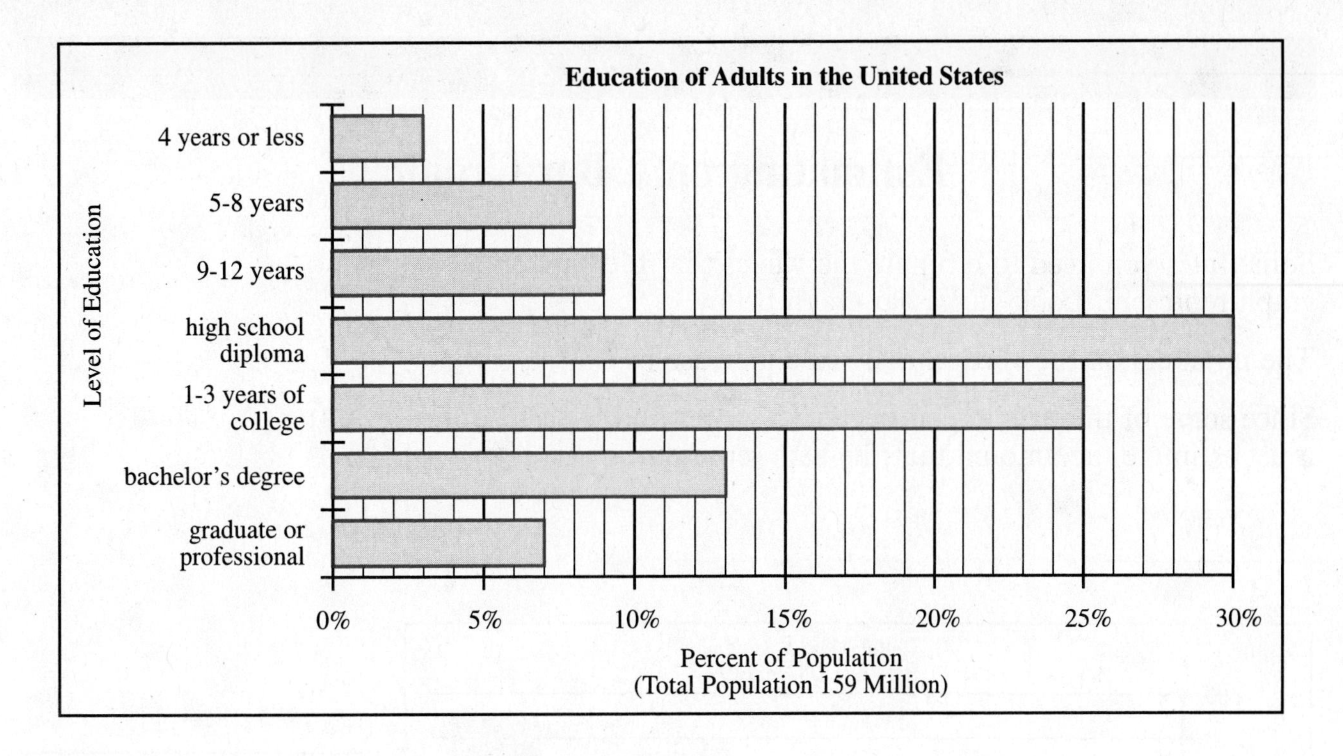

Use the bar graph above to answer questions 6–12.

6. What is the title of this bar graph? __________

7. What is the total population of adults? __________

8. How many categories of education are being displayed? __________

9. What percent of the total population has completed high school? __________

10. Which educational level has the lowest percentage? __________

11. Which educational level has the greatest percentage? __________

12. Which educational level(s) have percentages that are less than 10% of the population? __________

LESSON 32

Estimating on a Bar Graph

Sometimes you need to estimate the values that the bars on a bar graph represent. Look at the bar graph below.

The numbers on the vertical axis scale increase by 200.

Since some of the bars are between two values on the scale, you must estimate the amount that the bars represent.

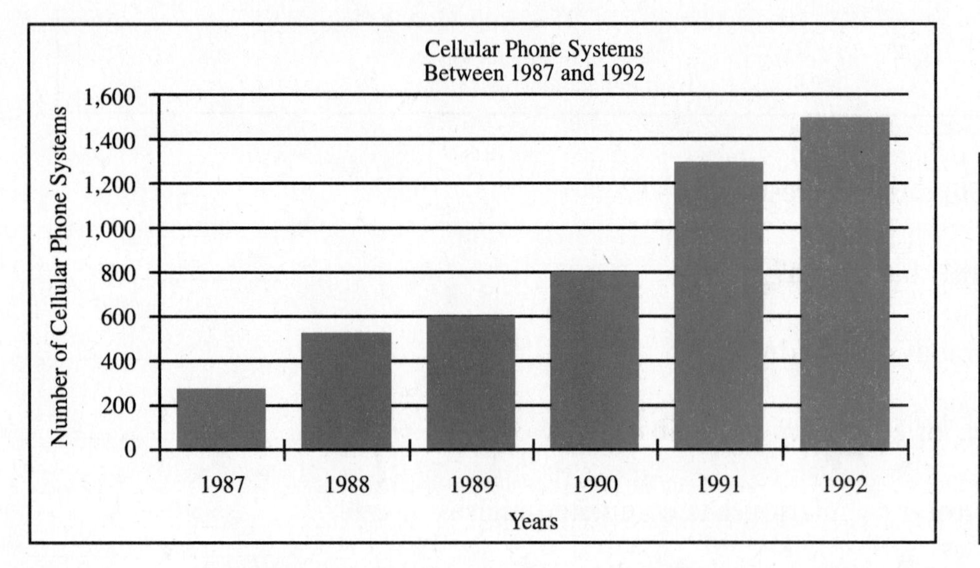

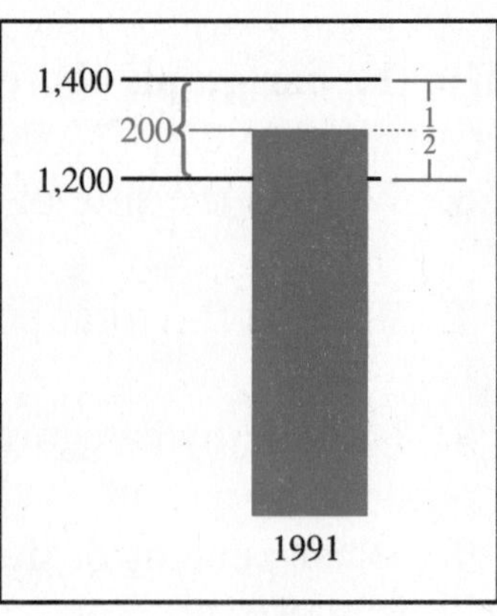

Example

The bar representing 1991 stops halfway between 1,200 and 1,400.

$\frac{1}{2}$ of 200 is 100.

100 + 1,200 is 1,300.

The number of cellular phone systems in 1991 is 1,300.

Practice

Use the bar graph above to answer questions 1–3.

How many cellular phone systems were there:

1. In 1987? __________

2. In 1988? __________

3. In 1992? __________

Use the bar graph below to answer questions 4–8.

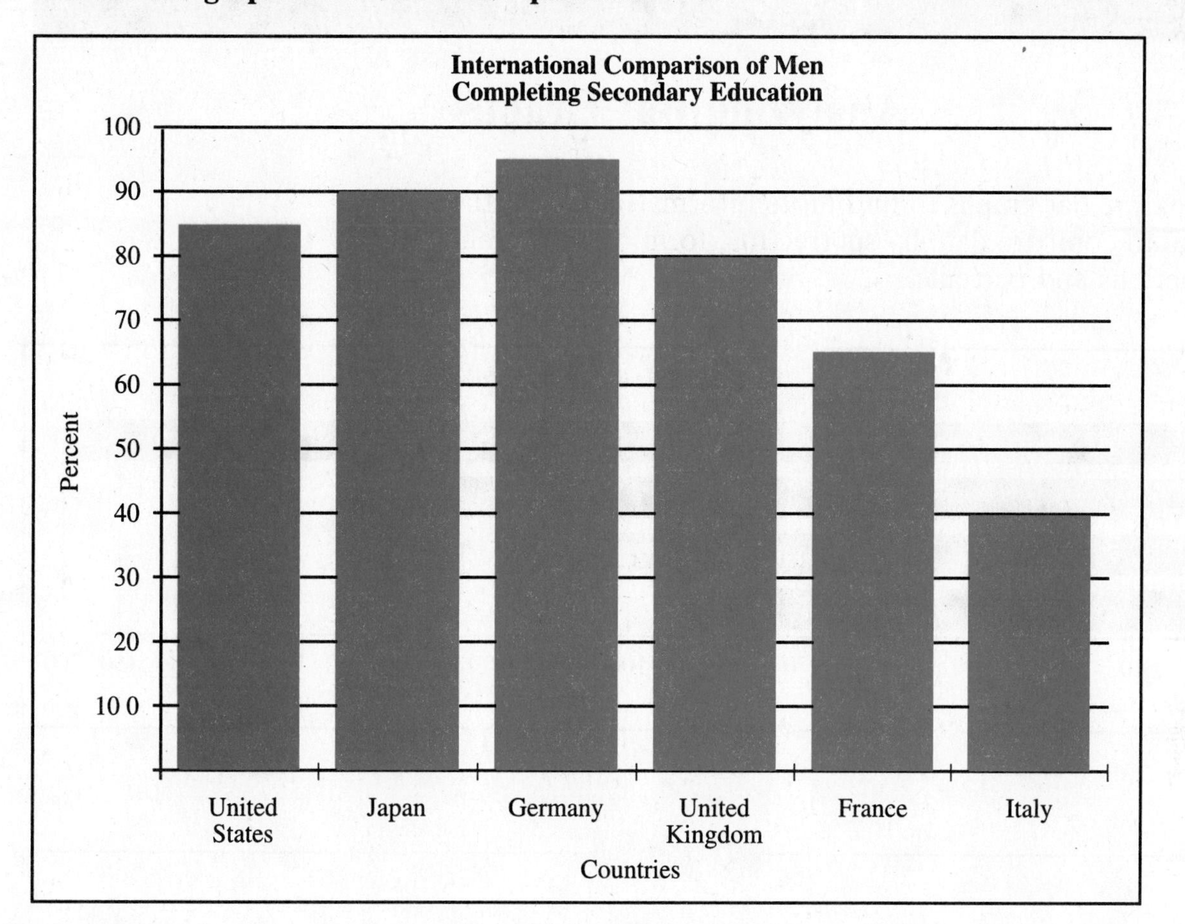

4. What is being described in this bar graph? ______________

5. Which country has the largest percentage of graduates? ______________

6. Which country has the lowest percentage of men completing secondary education? ______________

7. What percentage of men in Germany completed secondary education? ______________

8. What is the percentage of men in the United States who completed secondary education? ______________

LESSON 33

Analyzing Bar Graphs

You can analyze bar graphs to find more information about the data. It is helpful to compare data by subtracting, forming ratios, and finding fractions and percents.

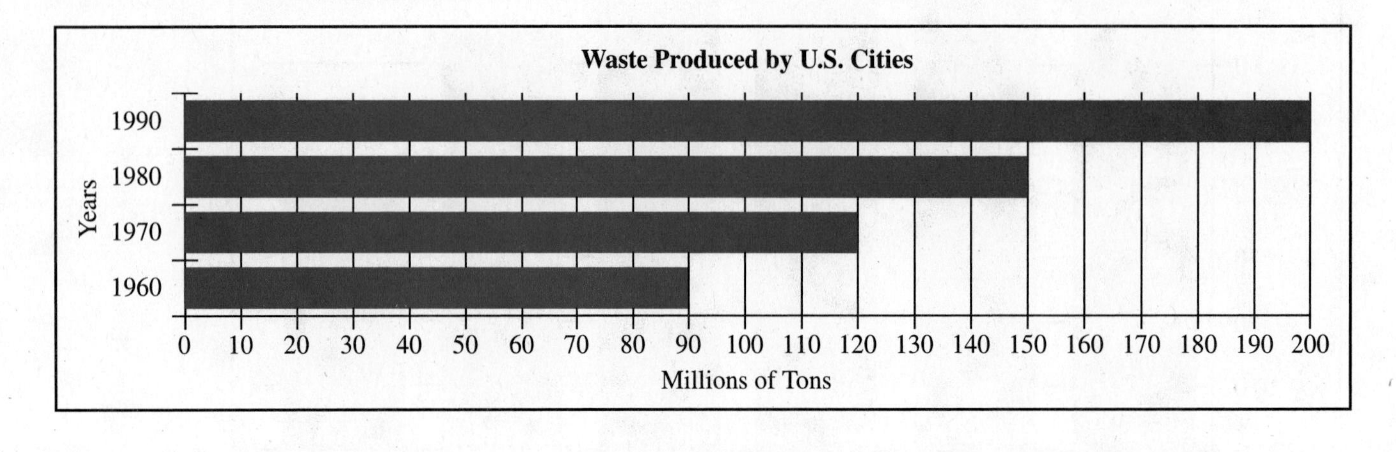

Example

Look at the bar graph above. The amount of waste produced in 1960 is what fraction of the waste produced in 1990?

The bar graph shows that 90 million tons of waste were produced in 1960 and that 200 million tons were produced in 1990.

$\frac{90 \text{ million}}{200 \text{ million}}$ Write this as a fraction.

$\frac{9}{20}$ Reduce.

The amount of waste produced in 1960 is $\frac{9}{20}$ of the amount produced in 1990.

Practice

Use the bar graph above to answer the questions 1–6.

1. How many millions of tons of waste were produced in 1970? ____________

2. The amount of waste produced in 1970 is how much more than the waste produced in 1960? ____________

3. How much less waste was produced in 1970 than in 1980? ____________

4. The amount of waste produced in 1960 is what fraction of the waste produced in 1980? ____________

5. What is the percent of increase in the waste produced from 1980 to 1990? ____________

6. The amount of waste produced in 1980 is what percent of the waste produced in 1990? ____________

Problem Solving

Use the graph on page 98 to solve the problems.

7. How many millions of tons of trash do you predict will be produced by U.S. cities in the year 2000? (Hint: Use the percentage of increase in millions of tons of trash between 1980 and 1990.) ____________

8. How can the information in the bar graph help cities plan for future waste removal? ____________

Use the bar graph below to answer questions 9–13.

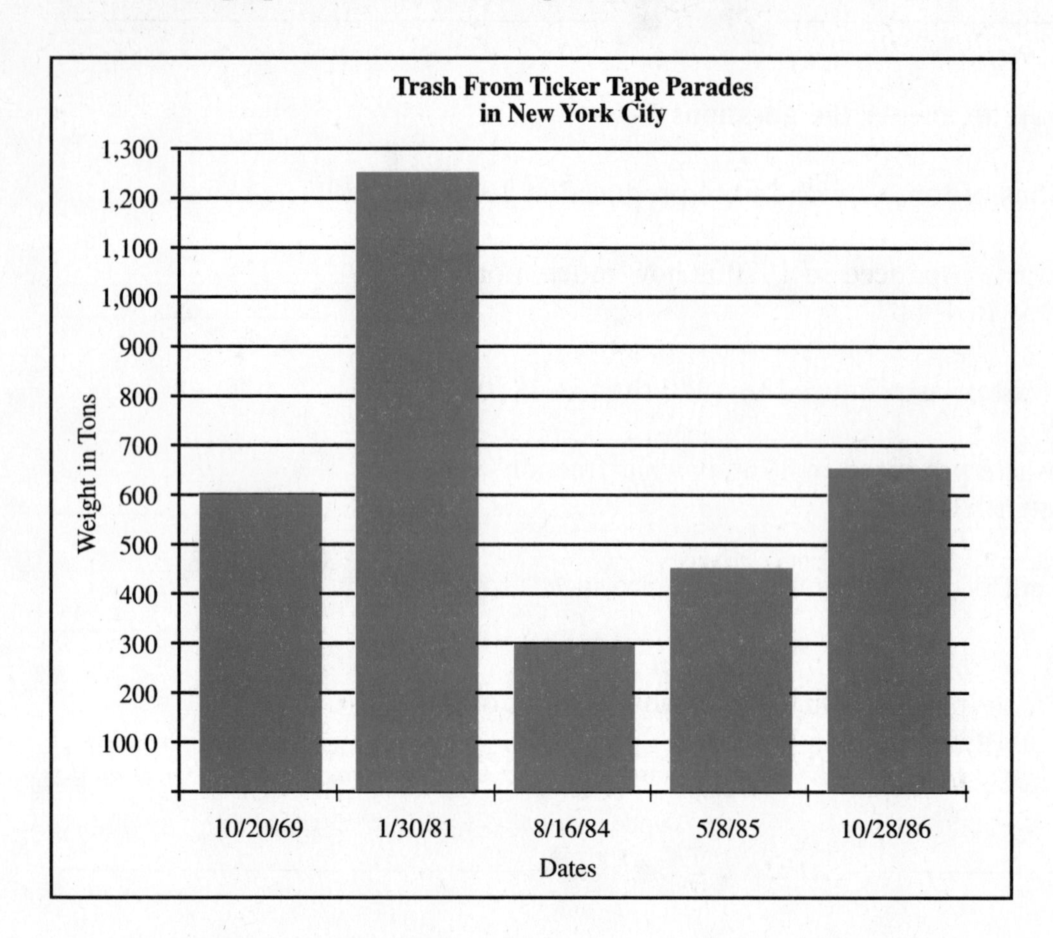

9. On January 30, 1981, the hostages were returned from Iran. The parade produced the largest amount of trash ever produced from a parade. How much trash was produced? __________

10. The least amount of trash was produced when the Olympic medalists were honored on August 16, 1984. How much trash was produced? __________

11. How much more trash was produced on October 28, 1986, when the New York Mets won the World Series than when they won it on October 20, 1969? __________

12. If the city can clean 50 tons of trash an hour, how many hours did it take for city workers to clean up after the parade on August 16, 1984? __________

13. If it cost $300 to clean up a ton of trash, how much did it cost to clean up after the Vietnam Veterans Salute on May 8, 1985? __________

LIFE SKILL

Reading an Electric Bill Bar Graph

The electric company is surveying customer energy usage. Below is a graph of the Navarro family's electric usage for one year.

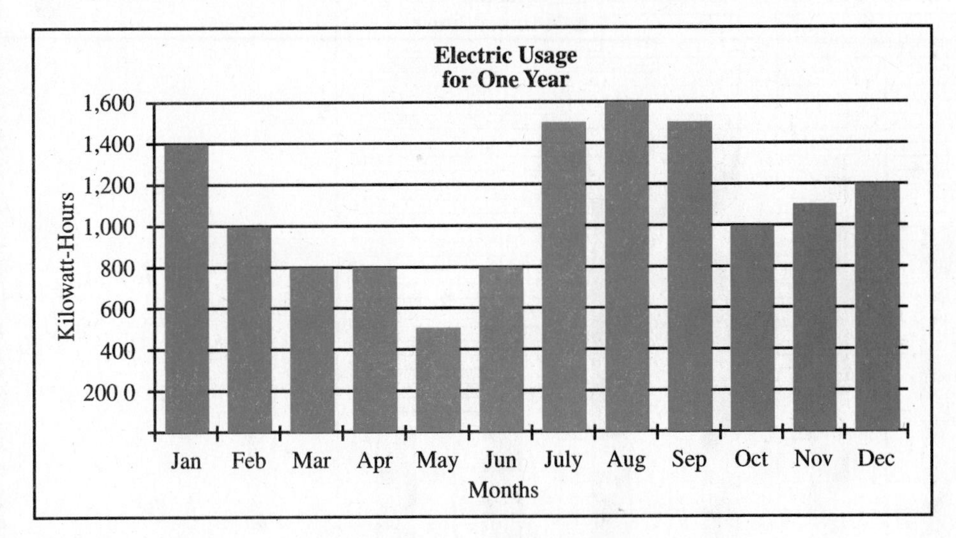

Use the graph above to solve the following problems.

1. How many more kilowatt-hours of electricity did the Navarros use in August than in May? __________

2. If the rate is $0.10 per kilowatt-hour, what is the cost of electricity in June? __________

3. How much more does the electric bill cost in September than in October? __________

4. During which four months did the Navarros use the most electricity? Why? __________

5. The electric company is offering a budget plan. The customers pay the same amount every month, based on their average monthly electric usage. According to the bar graph, how much would the Navarros pay each month on the budget plan? __________

LESSON 34

Double Bar Graphs

A **double bar graph** shows two sets of data at the same time. You can easily compare information and analyze data.

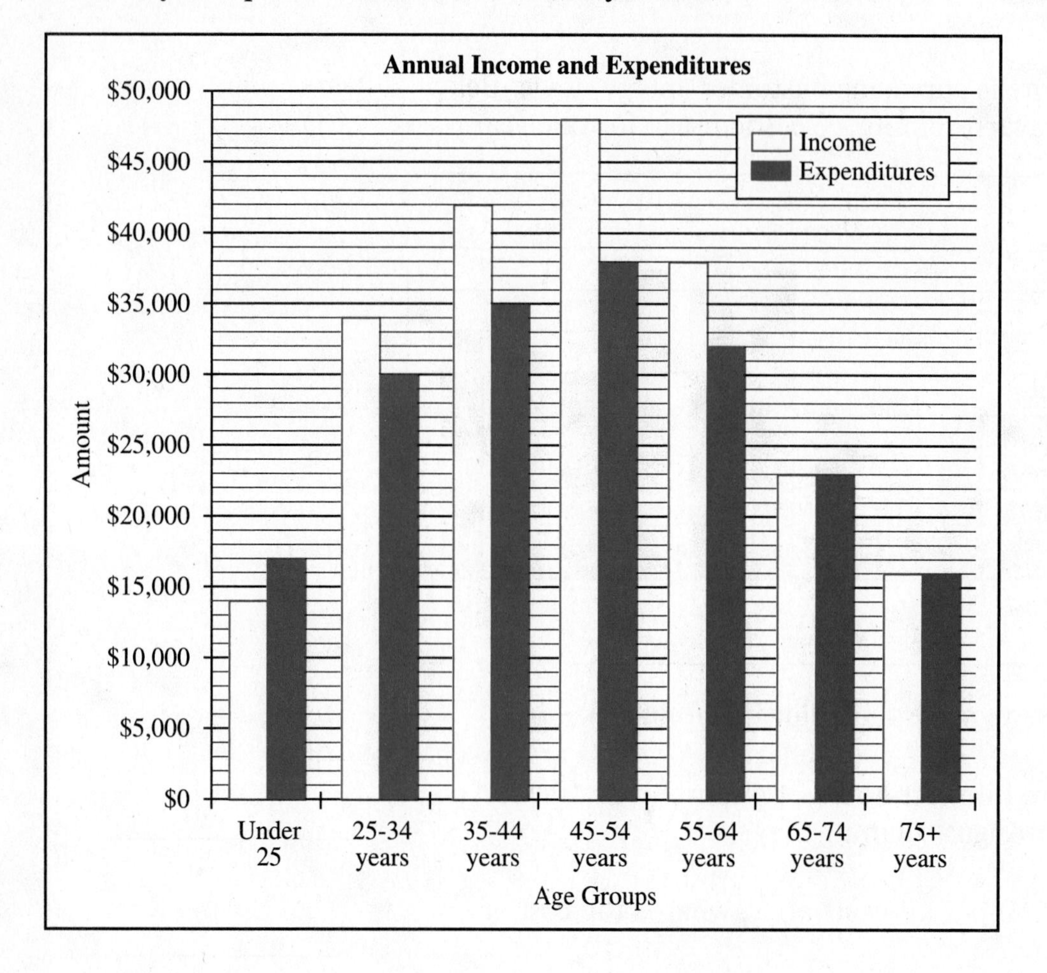

Example

A. Look at the double bar graph above. The left bar in each pair represents income. The right bar represents expenditures.

What is the difference between the income and the expenditures for people between the ages of 25 and 34?

Look to the pair of bars labeled "25–34 years."
The income bar represents $34,000.
The expenditure bar represents $30,000.

Subtract to find the difference.

$\$34{,}000 - \$30{,}000 = \$4{,}000$

The difference between income and expenditures for the 25–34 age group is $4,000.

Practice

Use the double bar graph on page 102 to answer questions 1–6.

1. In which age group do people spend more money than they earn?

   ____

   What are some reasons for this?

   ____

2. How much more money is earned per year than is expended by people in the group aged 55–64 years?

   ____

3. After the age of 44, what is the trend in the relationship between income and expenditures?

   ____

4. What happens between income and expenditures of people who are 65 and older?

   ____

   Why might this occur?

   ____

5. In which age group is the difference between income and expenditures the greatest?

   ____

   Why is the difference so great?

   ____

6. Expenditures are what percent of the annual income earned by people aged 35–44 years?

   ____

Use the double bar graph below to answer questions 7–13.

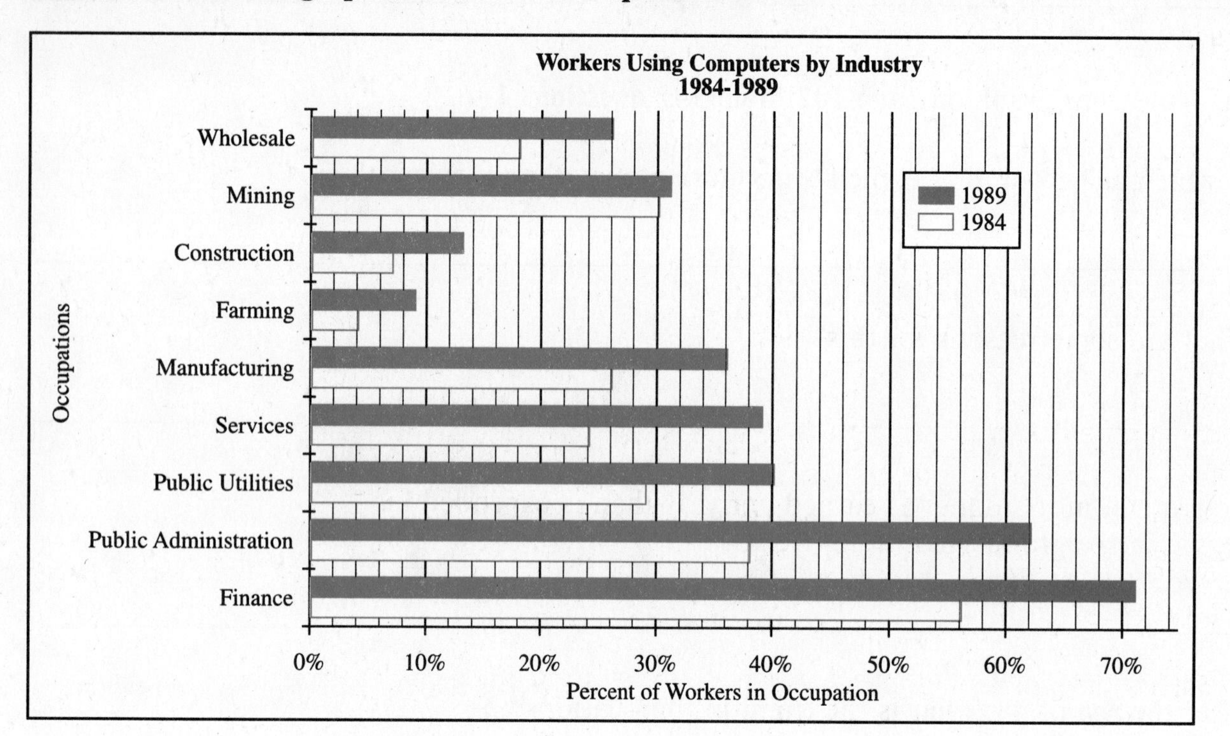

7. In the public utilities industry, what is the difference in the percent of computer use between the years 1984 and 1989? __________

8. Which industry had the greatest growth in the use of computers between 1984 and 1989? __________

9. Which industry had the least growth in the use of computers between 1984 and 1989? __________

10. If there were 8,020,000 people working in the finance industry in 1989, how many people used computers? __________

11. Based on the trends in the graph, which two industries probably will have the greatest percentage of workers using computers in the next 10 years? __________

12. Which two industries will have the smallest percentage of workers using computers in the next 10 years? __________

13. Although the farming industry had the smallest percent of workers using computers, it had the greatest percent of increase from 1984 to 1989. The percent of farmers using computers more than doubled. If the rate stays the same, what percent of the farmers probably will use computers in the year 1999? __________

The double bar graph below shows three kinds of information at once: import car sales, domestic car sales, and total car sales.

The whole bars represent total car sales.
The bottom portion of the bars represent domestic sales.
The top portion of the bars represent import sales.

MATH HINT

You can find import sales by subtracting the domestic sales from the total sales.

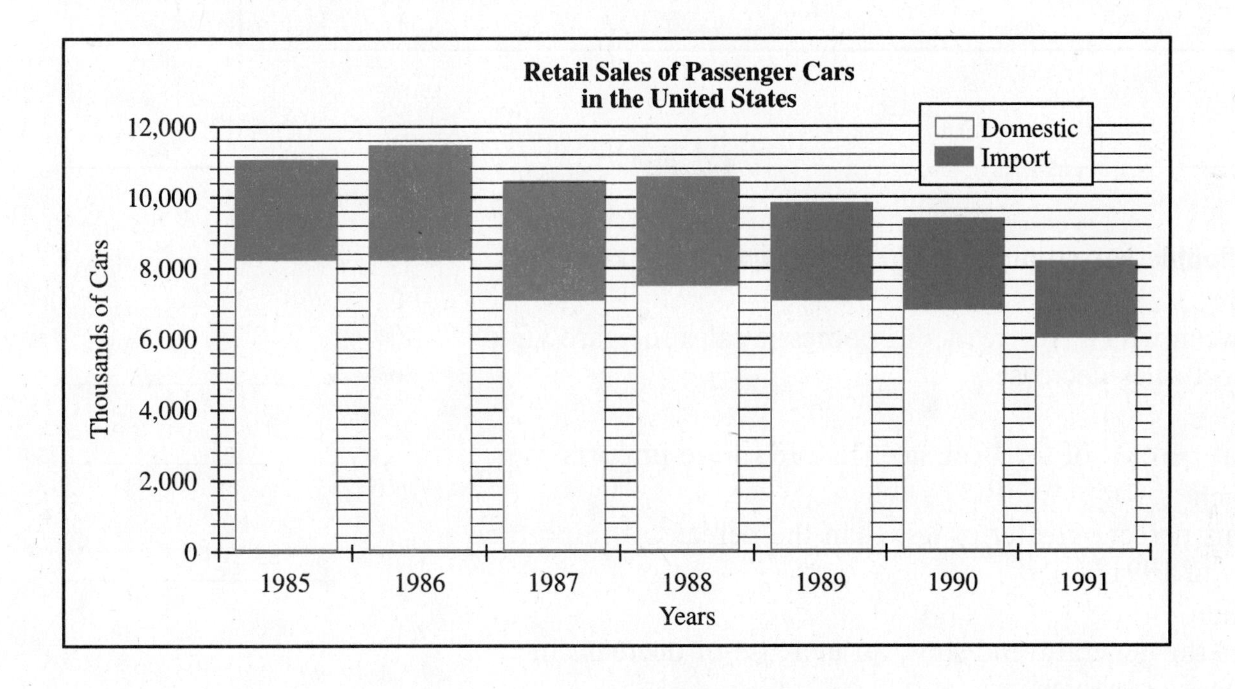

Example

B. What was the number of import car sales in 1985?
The total sales in 1985 was 11,000 (thousands).
The domestic sales in 1985 was 8,100 (thousands).

$11{,}000 - 8{,}100 = 2{,}900$ Subtract to find import car sales.

The number of import car sales in 1985 was 2,900 thousands, or 2,900,000.

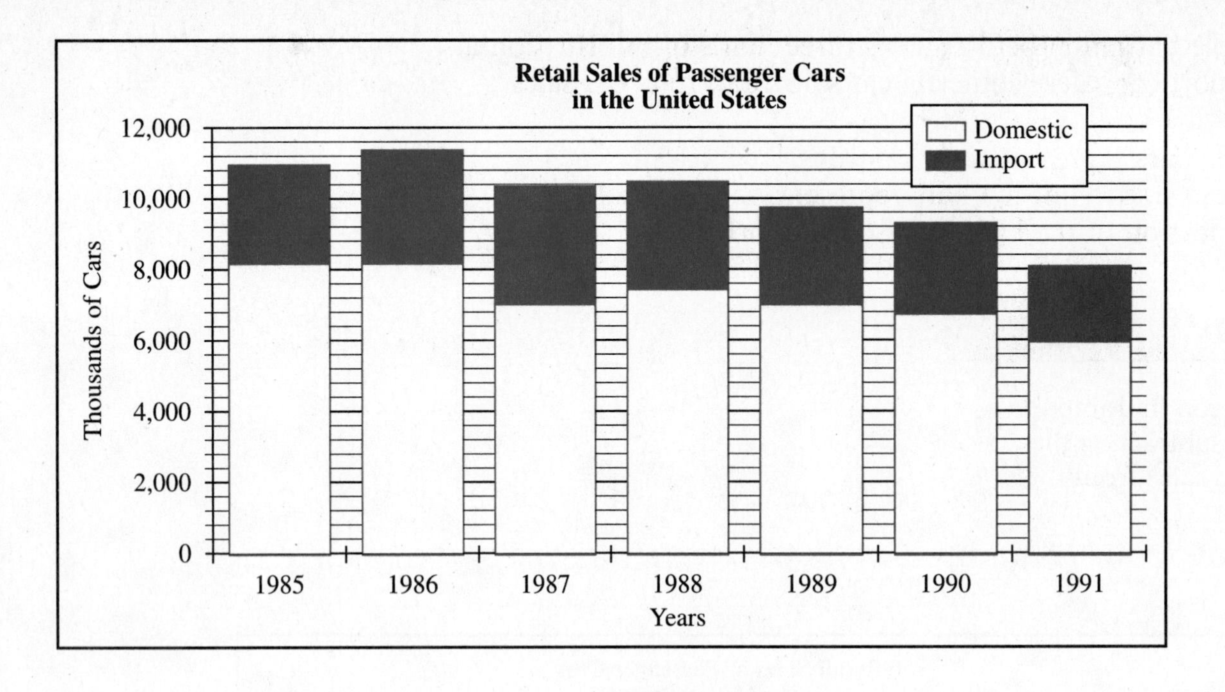

Practice

Use the double bar graph above to answer questions 14–19.

14. Between which two years did domestic sales increase and import sales decrease? ____________

15. What percent of total car sales in 1988 were imports? ____________

16. Is this percent greater or less than the percent of import car sales in 1991? ____________

17. Does the general trend show an increase or decrease in domestic car sales? ____________

18. What can you predict will happen to the relationship between import and domestic car sales during the next 20 years?

____

19. Are Americans buying more import cars or more domestic cars? Do you predict this trend will change? Why?

____

LESSON 35

Reading a Line Graph

A line graph is useful in showing trends or changes over a period of time.

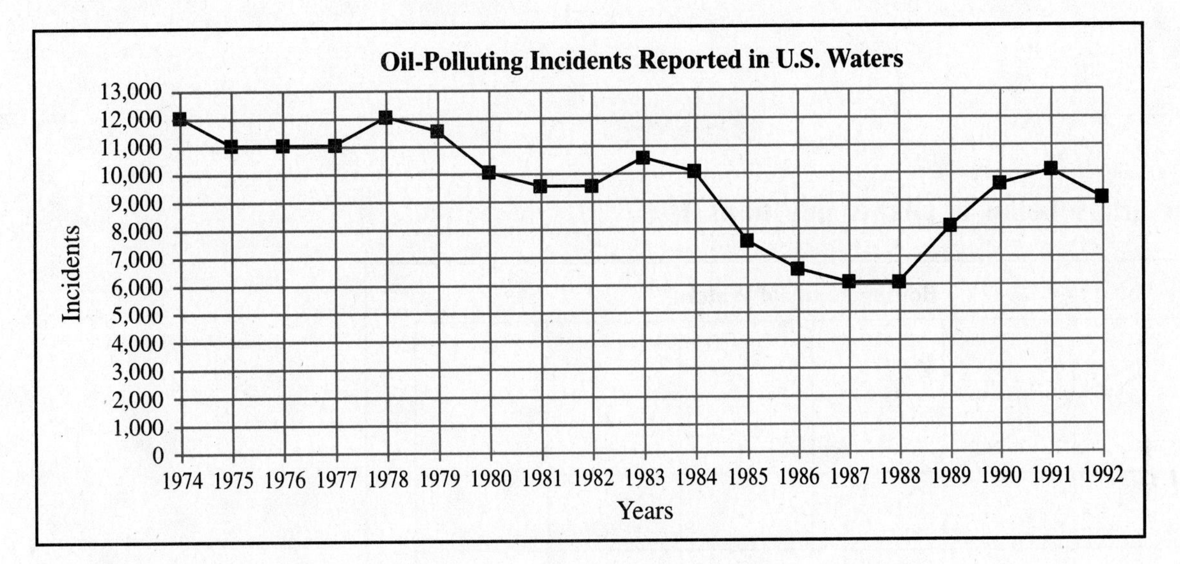

To read a line graph, keep in mind the following steps:

Step 1 Read the title of the graph and determine what is being compared.

Step 2 Read the labels of the horizontal and vertical axes to find what they represent.

Step 3 Determine the scale, or the units, used along the axes.

Step 4 Find a specific value by moving along the horizontal axis until you get to the desired unit. Go up to the line graph and then across to find the value on the vertical axis.

Example

How many oil polluting incidents occurred in 1987?

Step 1 The title of the graph is "Oil-Polluting Incidents Reported in U.S. Waters." The graph is comparing the number of oil-polluting incidents each year.

Step 2 The label of the horizontal axis is "Years" and the label of the vertical axis is "Incidents."

Step 3 Each unit on the horizontal axis represents one year and each unit on the vertical axis represents 1,000 oil-polluting incidents.

Step 4 Look along the horizontal axis until you get to the year 1987. Go up to the line graph and then across to the vertical axis. The value on the vertical axis is 6,000.

There were 6,000 oil-polluting incidents in 1987.

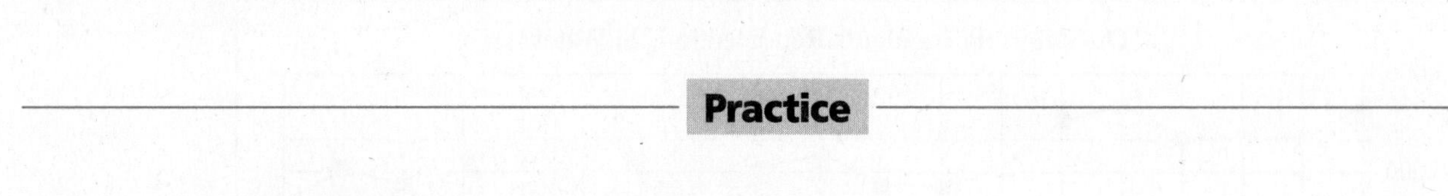

Practice

Use the line graph below to answer questions 1–6.

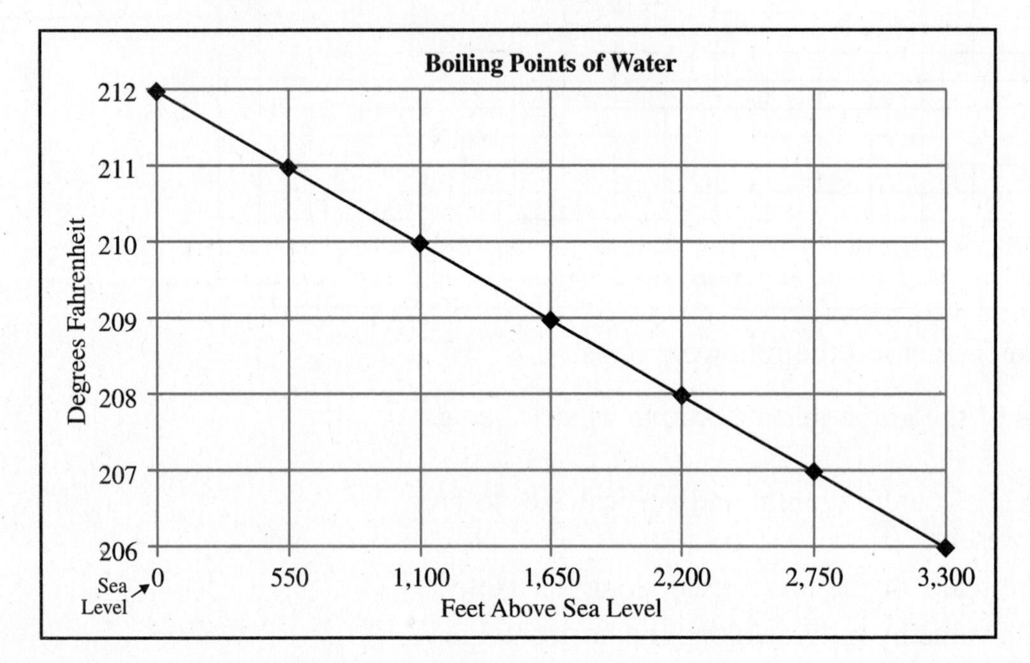

1. What is the title of the graph? __________
2. What is the label of the horizontal axis? __________
3. What is the label of the vertical axis? __________
4. What does each unit on the vertical axis represent? __________
5. At what temperature does water boil at 1,650 feet above sea level? __________
6. At what temperature does water boil at sea level? __________

LESSON 36

Analyzing Line Graphs

You can analyze line graphs by comparing the data in various ways. This is helpful in finding new information about the data. Below is a line graph displaying two sets of data at one time.

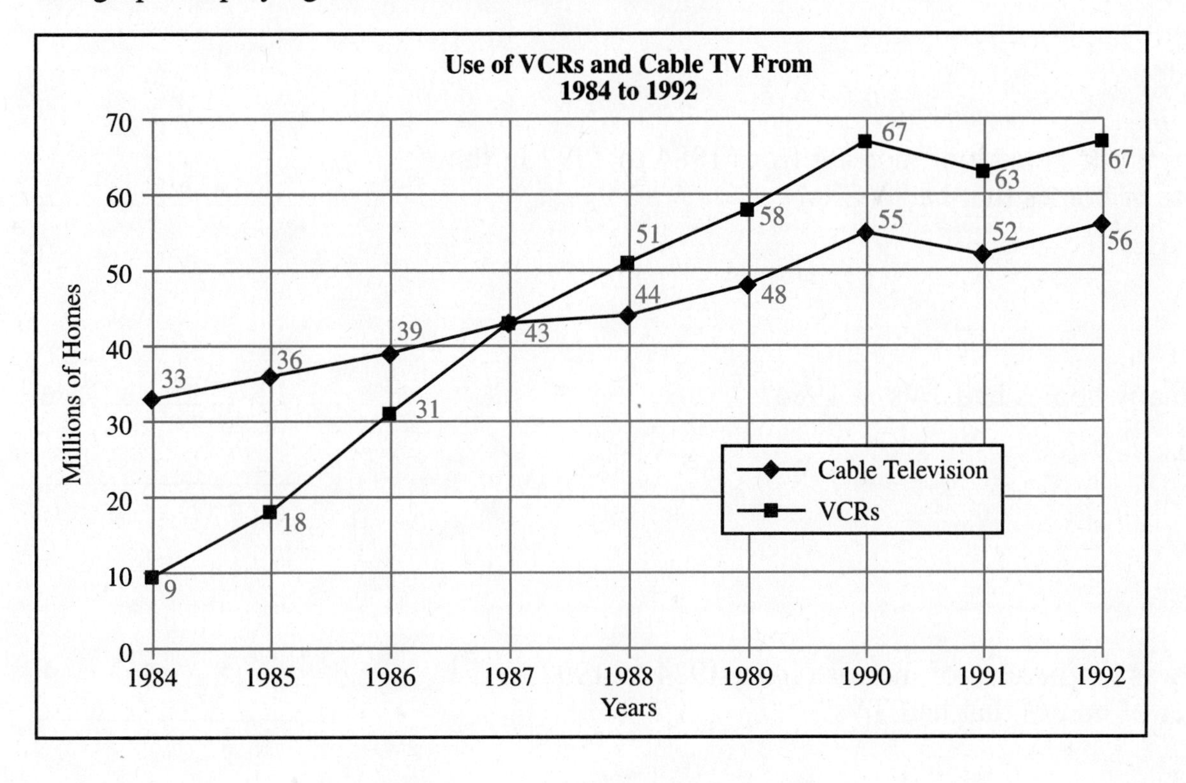

Example

Use the line graph to find how many more homes had VCRs than cable television in 1992.

Go to the year 1992 on the horizontal axis and find how many homes had VCRs and how many had cable television.

67 million had VCRs and 56 million had cable television.

$67 - 56 = 11$ Subtract to find the difference.

There were 11 million more homes that had VCRs than had cable TV in 1992.

Practice

Use the line graph on page 109 to answer questions 1–7.

1. How many homes had VCRs in 1984? __________

in 1992? __________

2. What was the percent of increase from 1984 to 1992 in the number of homes that had VCRs? __________

3. How many homes had TVs in 1984? __________

in 1992? __________

4. What was the percent of increase from 1984 to 1992 in the number of homes that had TVs? __________

5. Which is increasing at a faster rate, the number of homes with VCRs or the number with cable TV? __________

6. From 1984 to 1987, what is the trend in the relationship between the number of homes with VCRs and the number with cable TV? __________

7. If you were going to start your own business in 1998, would it be a cable repair business or video rental business? __________

Use the graph below to answer questions 8–13.

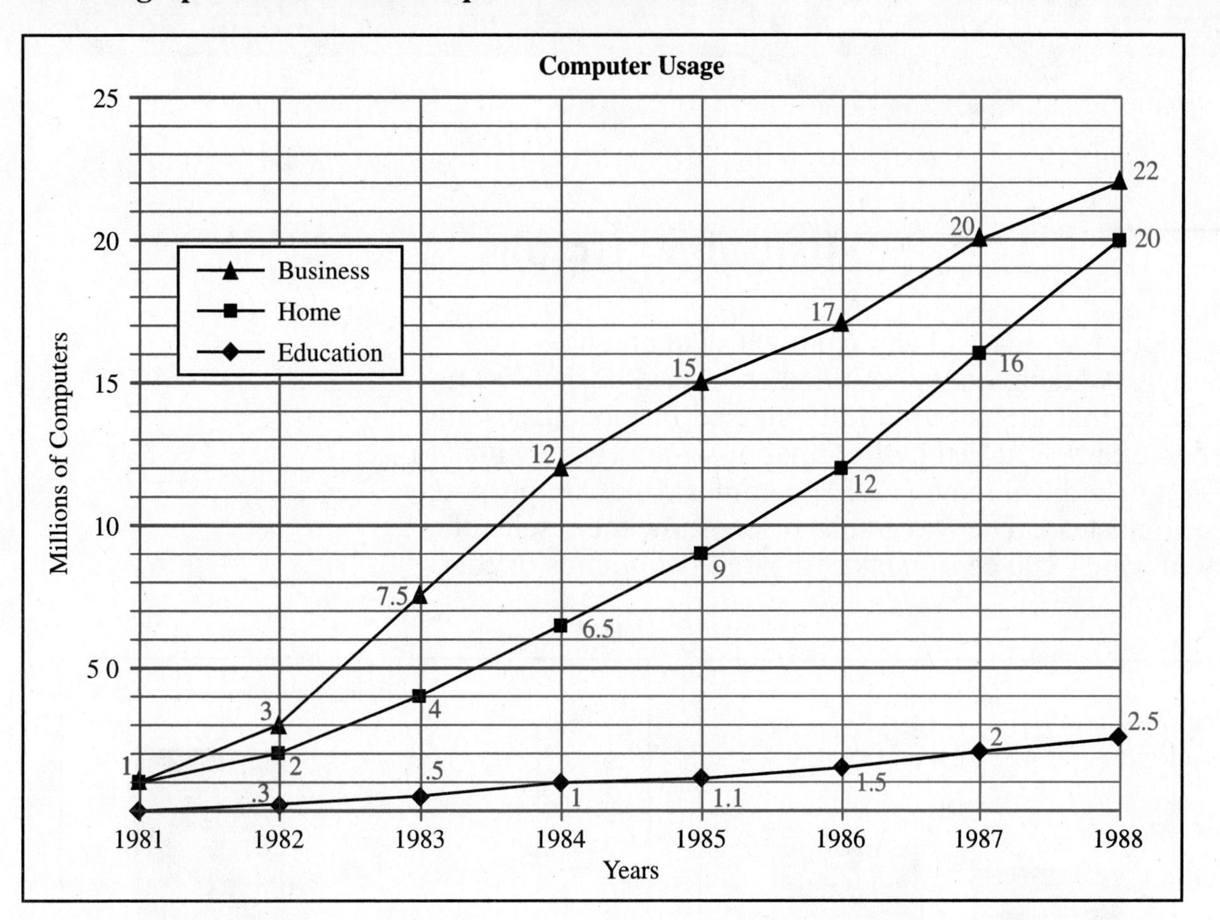

8. In 1988, how many times greater was the use of computers in business than in education? __________

9. What is the percent of increase in business use of computers between 1984 and 1985? __________

10. Which category shows the smallest increase in computer usage? __________

11. Do you think home computer usage will ever be greater than business computer usage? __________

    Give some reasons why this may or may not occur.

    ____

12. Predict the number of computers that will be used in business in the year 2000. __________

13. How will the trend in computer usage affect business growth?

    ____

LIFE SKILL

Misleading Graphs

Some graphs may inaccurately represent data and be misleading. A graph that is not titled or labeled or that has units in the scales that are not uniformly marked may be misleading.

People can be misled by information in advertisements or television or radio commercials. Sometimes scales are purposely chosen to mislead. The two graphs below show the results of a survey in which 100 consumers compared two brands of cola.

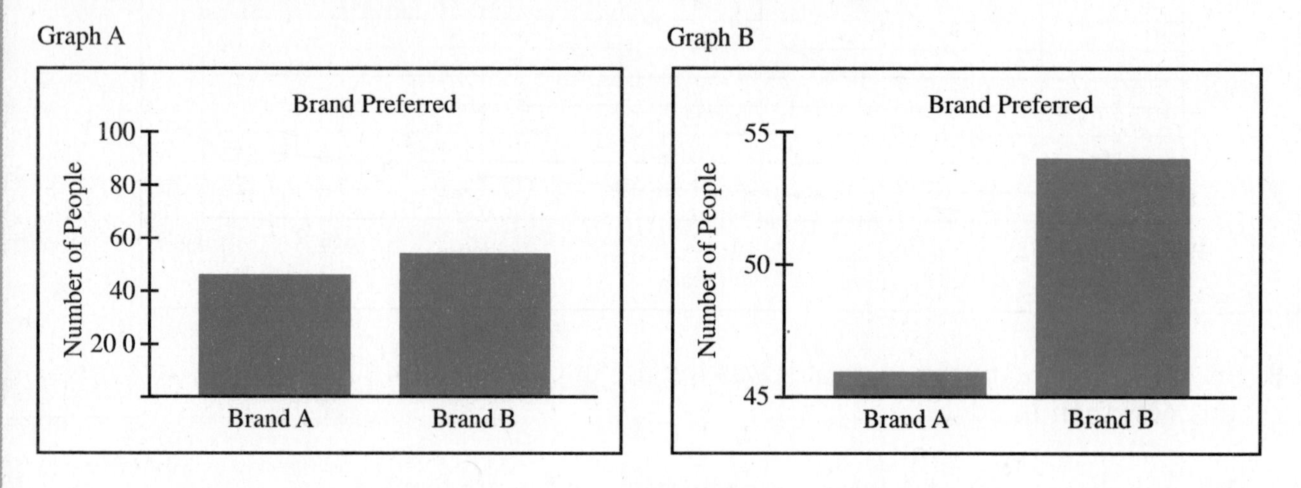

Use Graphs A and B to answer questions 1–3.

1. Which graph is misleading, Graph A or Graph B?

____

2. Why is the graph misleading?

____

3. Which graph would the company that sells Brand B cola want to use in advertising?

____

The two graphs below display the same monthly sales information for Alicia's Boutique. The letters on the horizontal axis represent the months, beginning with January.

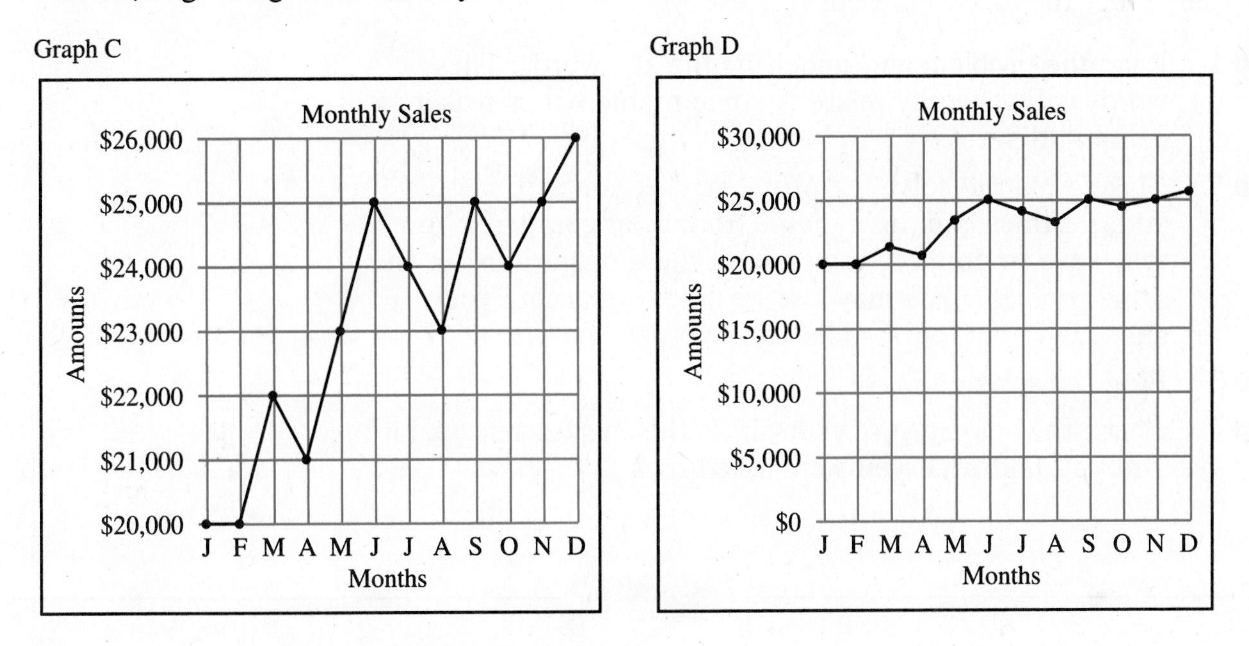

Use Graphs C and D to answer questions 4–6.

4. Is Graph C misleading? If so, why? ________________

____

5. Which graph would Alicia show to someone who wanted to buy her boutique? Why? ________________

____

6. Which graph would she show to an employee who asked for a big raise? Why? ________________

____

LESSON 37

Problem Solving—Using Bar Graphs and Line Graphs

You can solve word problems by using bar graphs and line graphs. It is sometimes easier to interpret information that you need to solve a problem when it is displayed in a graph.

Remember the following steps to solve word problems:

Step 1 Read the problem and underline the key words. These words will generally relate to some mathematics reasoning computation.

Step 2 Make a plan to solve the problem. Ask yourself, Should I add, subtract, multiply, divide, round, or compare? You may have to do more than one of these operations for the same problem. You may also be able to estimate your answer.

Step 3 Find the solution.

Step 4 Check the answer. Ask yourself, Is this answer reasonable? Did you find what you were asked for?

Example

Gwen owns stocks in Alpha Computers and in B&B Toys. On the first day of every month she graphs the closing values of a share of each stock. The graph on the right shows the values of her stocks for one year.

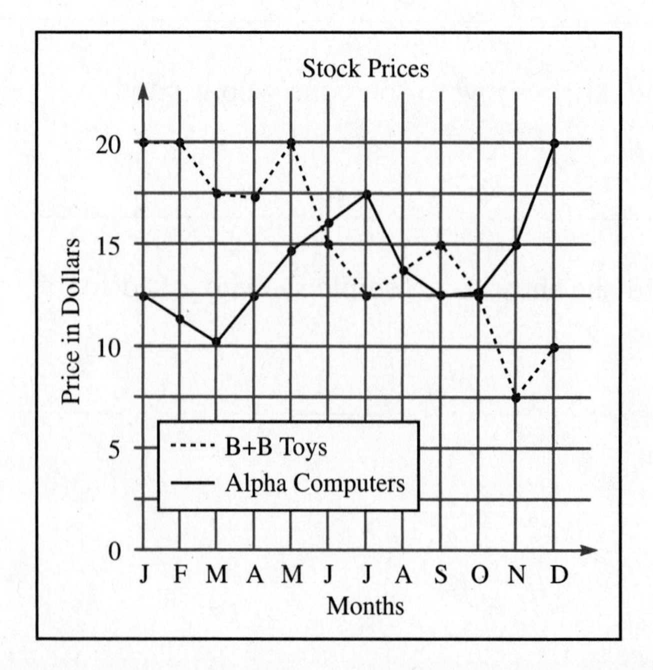

In May, how much greater was the value of the B&B Toys stock than that of Alpha Computer stock?

Step 1 Determine how much greater B&B Toys stock was worth than Alpha Computer stock. The key words are **how much greater.**

Step 2 The key words indicate which operation should occur—subtraction.

Step 3 Find the solution. Look at the graph to determine the value of each of the stocks in May. Alpha Computers stock was worth \$15 per share and B&B Toys stock was worth \$20 per share.

$\$20 - \$15 = \$5$ Subtract to find the difference.

In May B&B Toys stock was \$5 greater per share than the Alpha Computers stock.

Step 4 Check the answer. Does it make sense that B&B Toys stock was worth \$5 more than Alpha Computers stock in the month of May? Yes, the answer is reasonable.

Practice

Use the graph on page 114 to solve problems 1–4.

1. What is the overall trend in the B&B Toys stock? ______

2. What is the relationship between the two stocks from April to May? ______

May to July? ______

3. If Gwen keeps both stocks for another year, which stock do you think will be more profitable? ______

4. Do you think Gwen should continue to keep the stocks or sell one or both of them? Why? ______

______

Use the graph and information below to solve problems 5–10.

For a school poll, Tricia asked 25 people what they preferred to watch on television. Four people preferred music videos, 8 preferred comedy programs, 3 preferred the news, 4 preferred movies, and 6 preferred to watch drama series. Tricia constructed a bar graph to help her analyze the data.

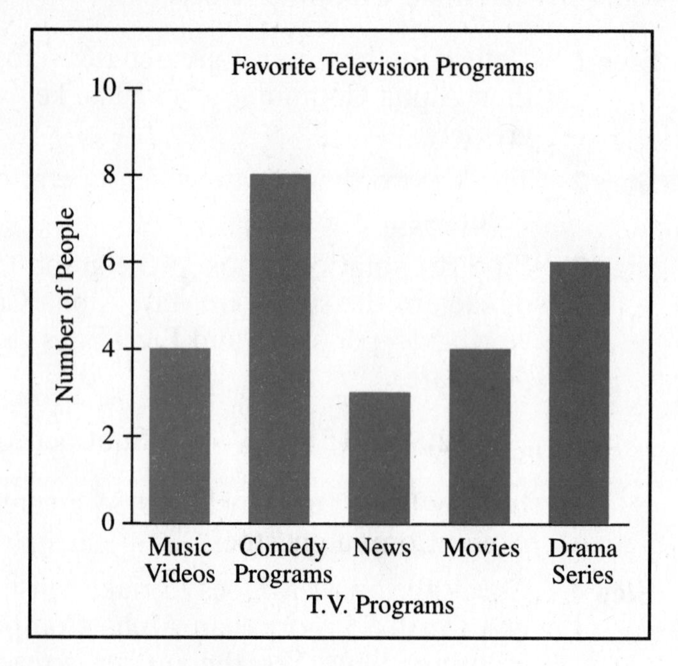

5. How many more people prefer to watch a drama series rather than the news? __________

6. The number of people who prefer to watch music videos is what percent of the number of people who watch comedy programs? __________

7. Which two types of programs had exactly the same number of people who prefer them? __________

8. According to this data, if there were 100 people, how many would prefer to watch movies on television? __________

9. If you were a product advertiser, during which type of program would you want your commercial to appear? __________

10. If there are 1,600 students in Tricia's school, how many would prefer to watch a drama series? __________

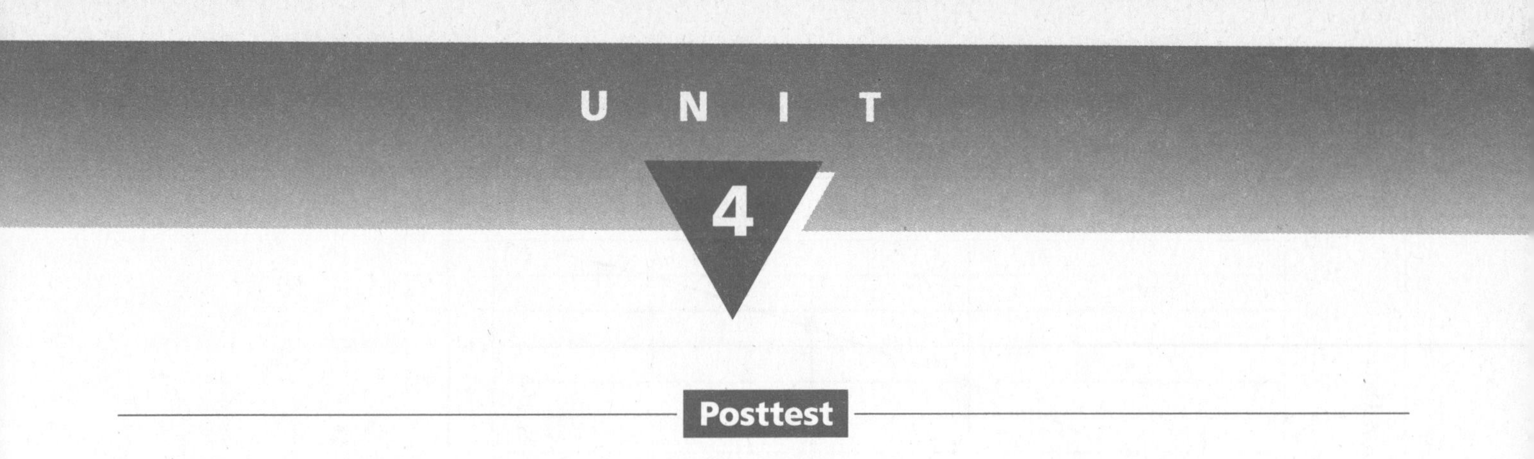

Posttest

Refer to the graph below to answer questions 1–4.

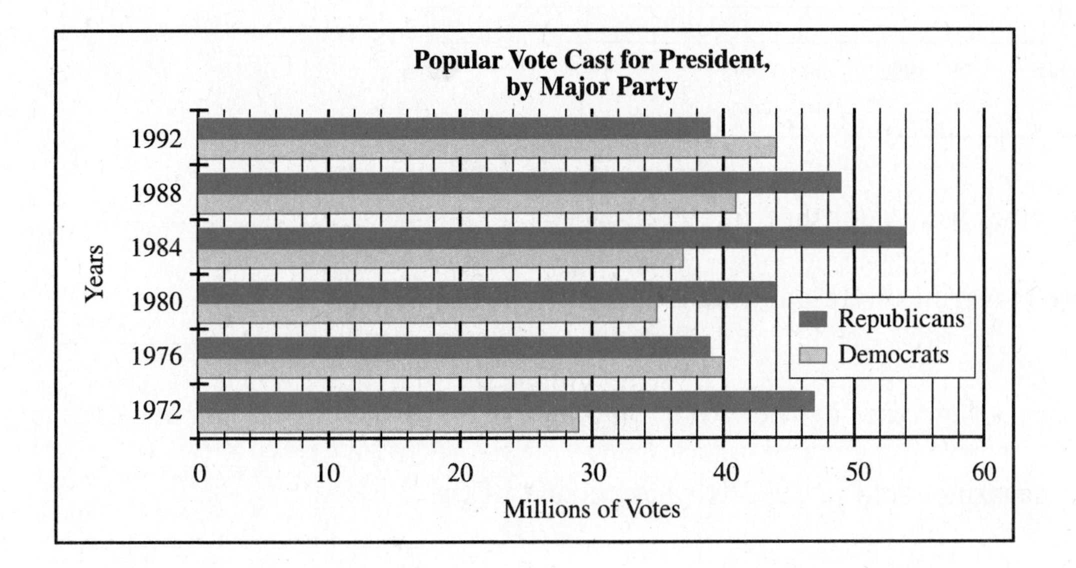

1. In 1992, the democratic candidate won by how many votes? ____________

2. In 1984, how many more votes did the Republican candidate get than the Democratic candidate? ____________

3. What was the total number of votes cast for both candidates in 1972? ____________

4. The votes cast for the Democratic candidate is what fraction of the total votes cast for both candidates in the 1988 election? ____________

Refer to the graph below to answer questions 5–8.

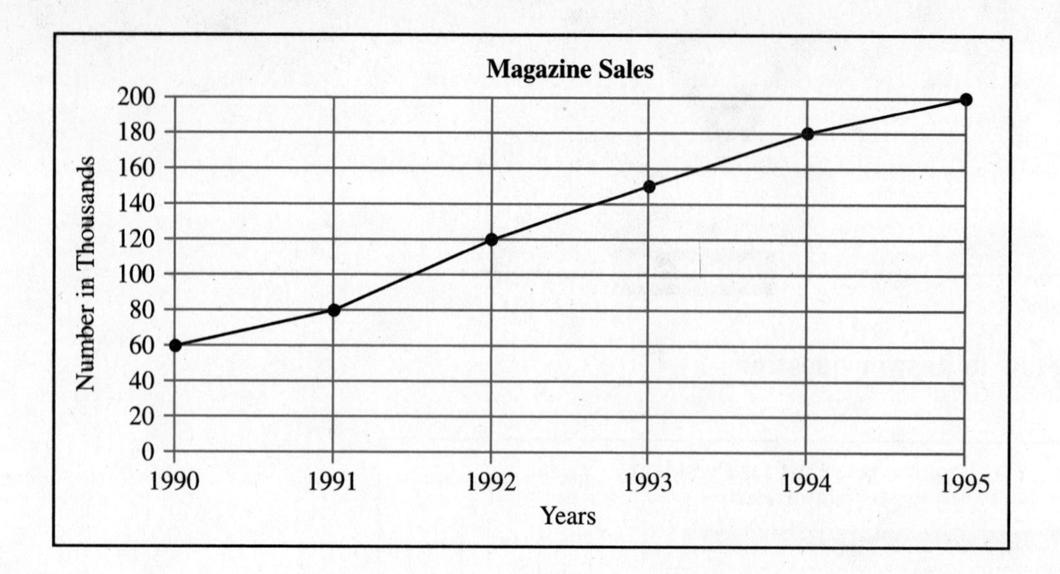

5. How many magazines were published in 1992? ____________

6. How many more magazines were published in 1993 than in 1990? ____________

7. What is the percent of increase in sales from 1990 to 1994? ____________

8. The number of magazines sold in 1992 is what percent of the number sold in 1995? ____________

Refer to the graph below to answer questions 9–14.

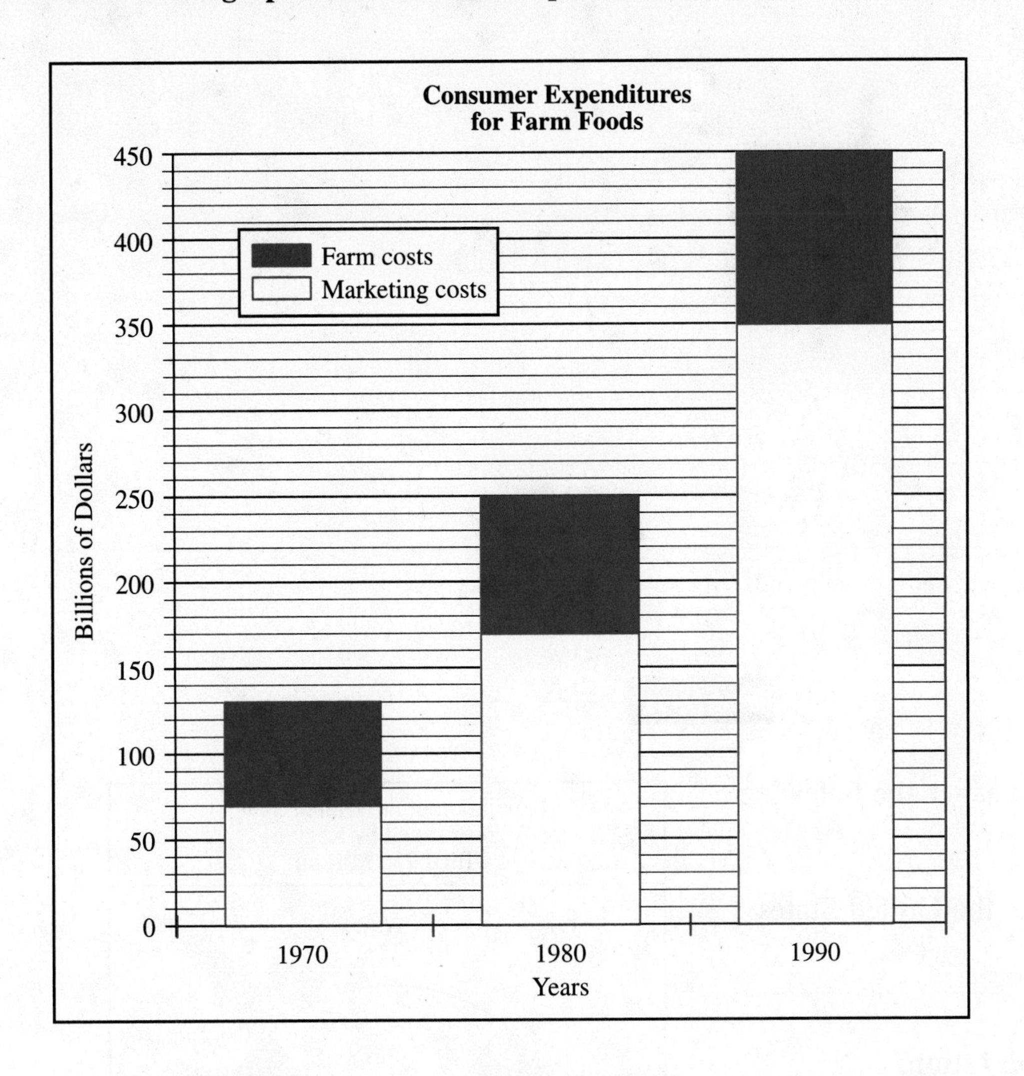

9. What is the difference in the total cost of food between 1980 and 1990? ____________

10. How many times greater are the marketing costs in 1990 than in 1970? ____________

11. What is the difference between the farm costs and the marketing costs for food in 1970? ____________

12. The farm costs of food is what percent of the total costs in 1980? ____________

13. Which costs are rising at a faster rate? ____________

14. By the year 2000, what percent of the total costs do you think marketing costs will be? ____________

UNIT

Circle Graphs

Pretest

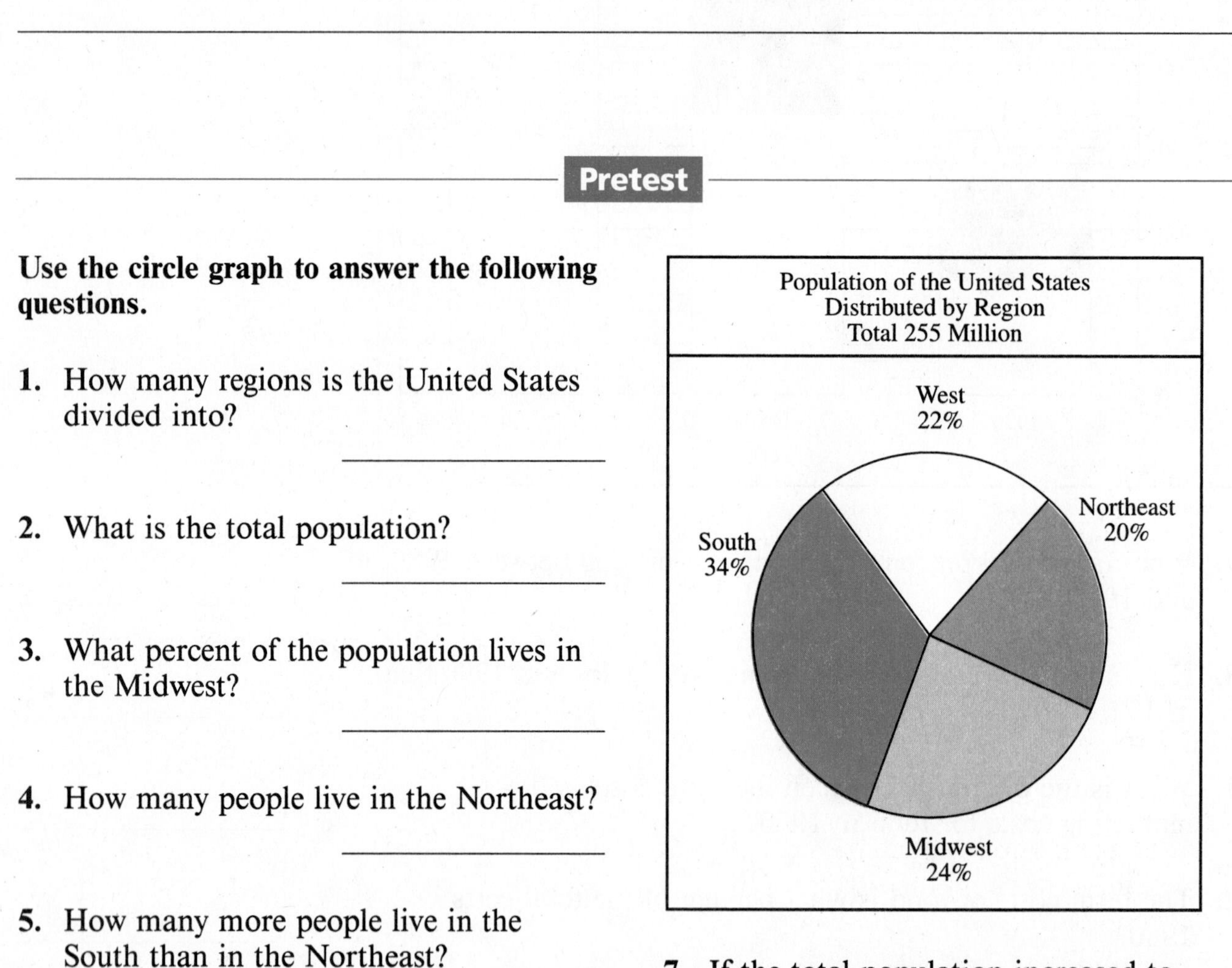

Use the circle graph to answer the following questions.

1. How many regions is the United States divided into? __________

2. What is the total population? __________

3. What percent of the population lives in the Midwest? __________

4. How many people live in the Northeast? __________

5. How many more people live in the South than in the Northeast? __________

6. How many people live in the West? __________

7. If the total population increased to 300,000,000, and the percentage remained the same, how many people would live in the West? __________

Reading a Circle Graph

A **circle graph** shows how parts of data are related to the whole. The entire circle represents the whole, and each section of the circle corresponds to a part of the whole. Percents are often used to represent parts of the whole.

> **MATH HINT**
>
> All the parts of a circle graph will add up to 100%.

Example

What percent of the GED credentials are issued to people 30–34 years old?

Find the part of the circle labeled "30–34" and see what percentage it is. It is 10%, so 10% of the GED credentials are issued to people 30–34 years old.

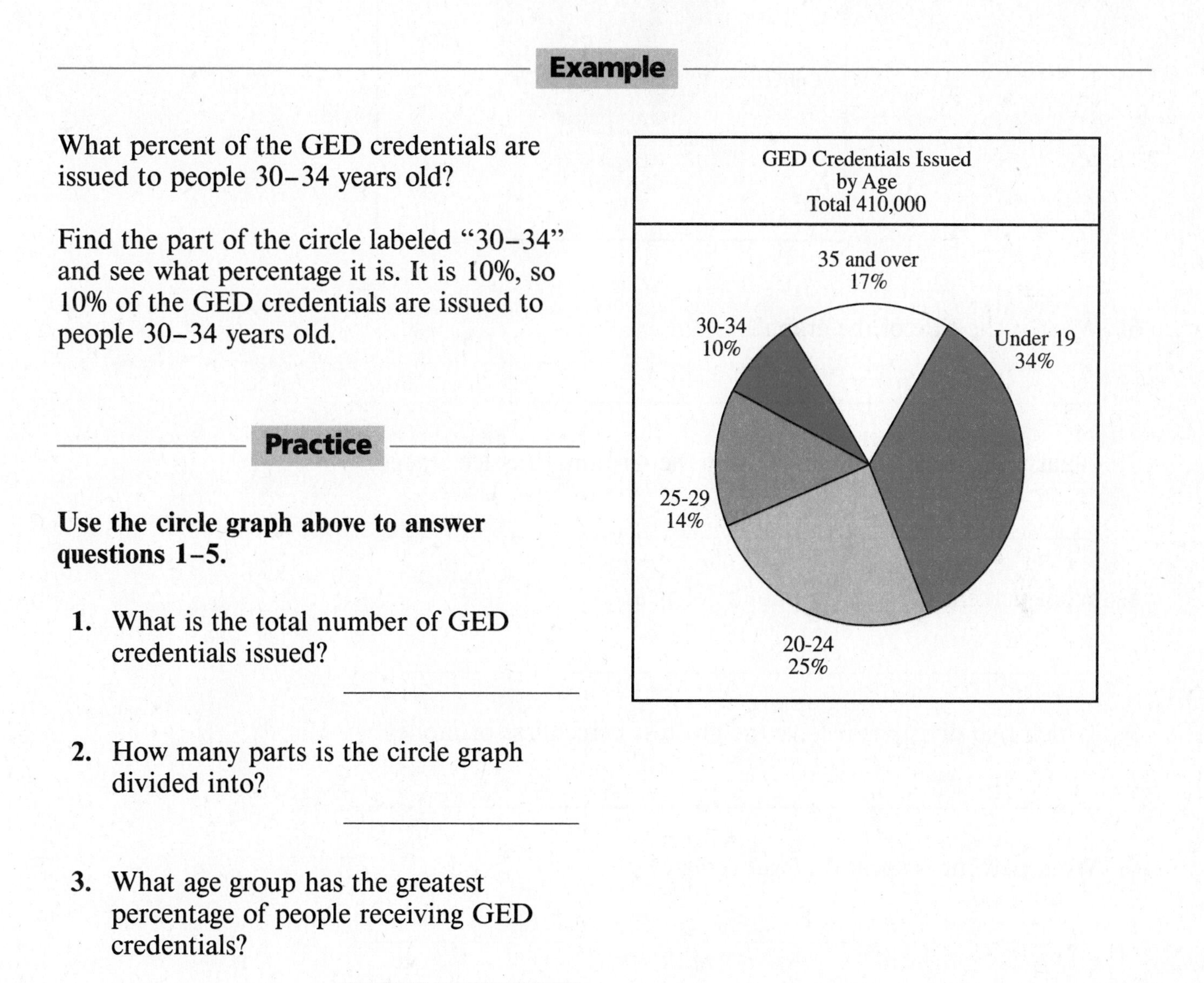

Practice

Use the circle graph above to answer questions 1–5.

1. What is the total number of GED credentials issued?

   ____

2. How many parts is the circle graph divided into?

   ____

3. What age group has the greatest percentage of people receiving GED credentials?

   ____

4. What percent of the GED credentials is issued to people aged 20–24?

   ____

5. What percent of the GED credentials is issued to people 35 and over?

   ____

Use the graph below to answer questions 6–10.

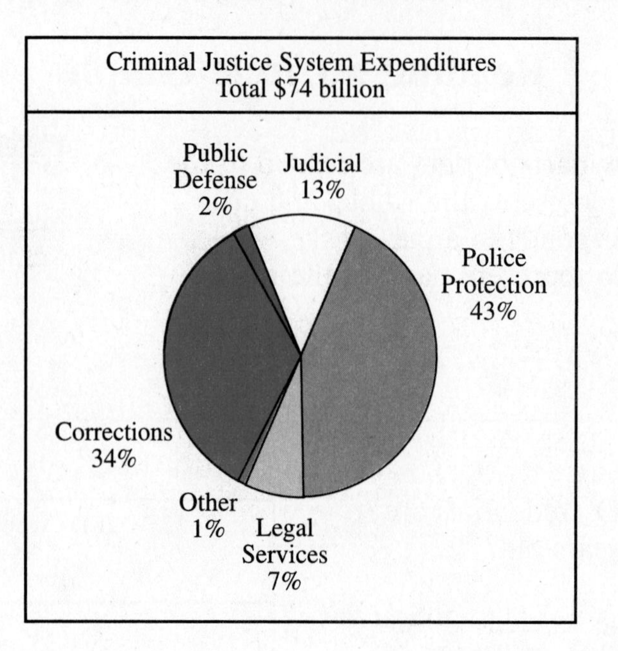

6. What is the title of the graph?

____

7. What is the total amount spent by the Criminal Justice System?

____

8. What percent is spent on public defense?

____

9. Which two divisions receive the greatest percentage of money?

____

10. What percent is spent on legal services?

____

LESSON 39

Analyzing a Circle Graph

In a circle graph, the circle represents 100% of the total data. It is helpful to analyze a circle graph to find information about the individual parts of the data and how they relate to the whole.

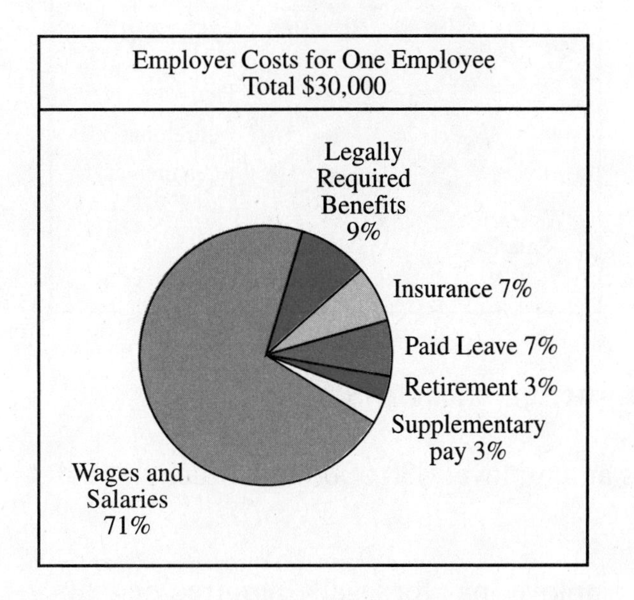

Example

The employer decides to increase the percent paid for insurance and decrease the wages of the employee to $20,400. How will the percentages of the employer's costs change?

Set up a proportion to find what percent of the total costs is wages. $\frac{20{,}400}{30{,}000} = \frac{n}{100}$

Use your math knowledge to solve for n. $n = \frac{20{,}400 \times 100}{30{,}000}$

The decreased wage makes up 68% of total employer costs. $n = 68\%$

The new percentage is 3% less than the original percentage for wages and salaries. Since all parts of the circle graph must equal 100%, the percent paid for insurance will be 3% greater, or 7% + 3% = 10%.

Wages and salaries account for 68% of total employer costs; insurance accounts for 10% of employer costs.

Practice

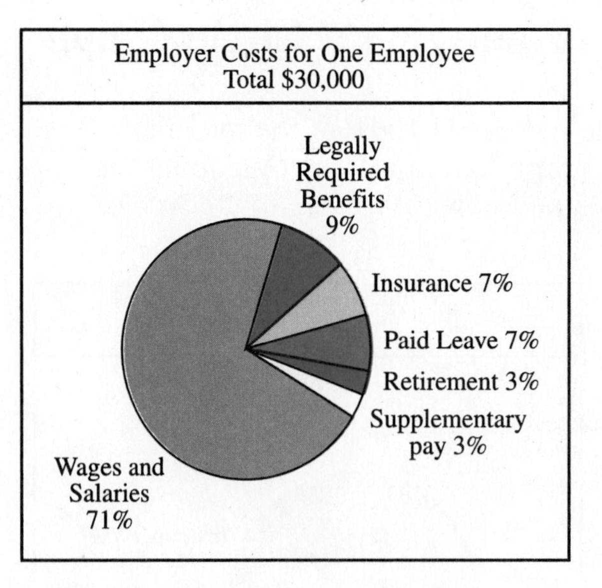

Use the graph above to answer questions 1–5.

1. If an employee costs an employer $30,000, how much is paid for wages and salary? __________

2. How much does an employer pay for legally required benefits for one employee? __________

3. What are the total costs for retirement and supplementary pay for one employee? __________

4. How much of the total costs is not wages and salary? __________

5. A company has four employees. Their total cost to the company is $90,000. Using the graph, how much does the company spend on legally required benefits for these employees? __________

Use the graph below to answer questions 6–11.

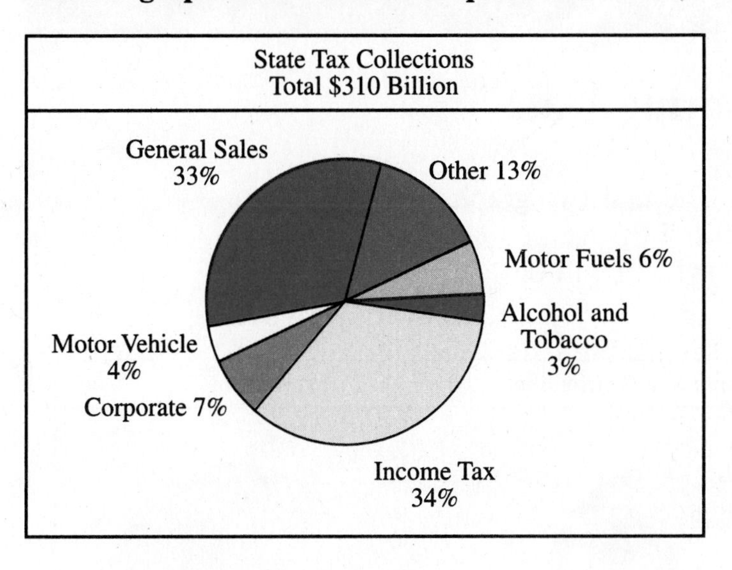

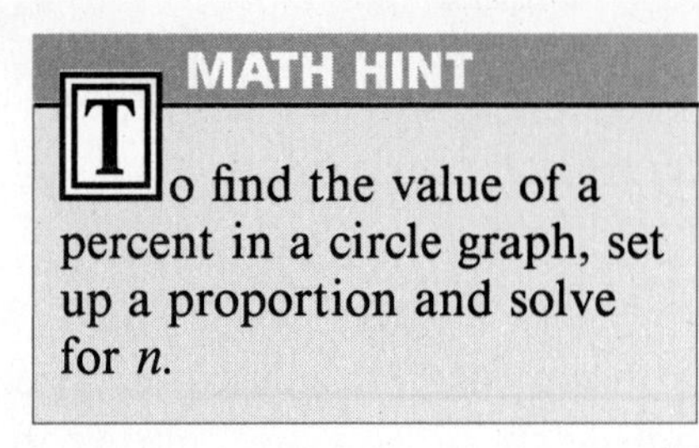

To find the value of a percent in a circle graph, set up a proportion and solve for *n*.

6. How many billions of dollars of state taxes come from individual income taxes? ______________

7. How much money comes from taxes on motor fuels? ______________

8. General sales revenues is how many times greater than taxes on alcohol and tobacco? ______________

9. How much more money is collected by states from taxes on motor fuel than from taxes on motor vehicles? ______________

10. If the total amount of state taxes collected was $400 billion, how much money would come from corporations? ______________

11. If a state wanted to increase its revenue by raising taxes, which taxable category would generate the most money?

____

LESSON 40

Comparing Data

You can use circle graphs to compare data. The first two graphs below show how the labor force was distributed by age in 1975 and 1990. The third graph predicts how the labor force will be distributed in 2005.

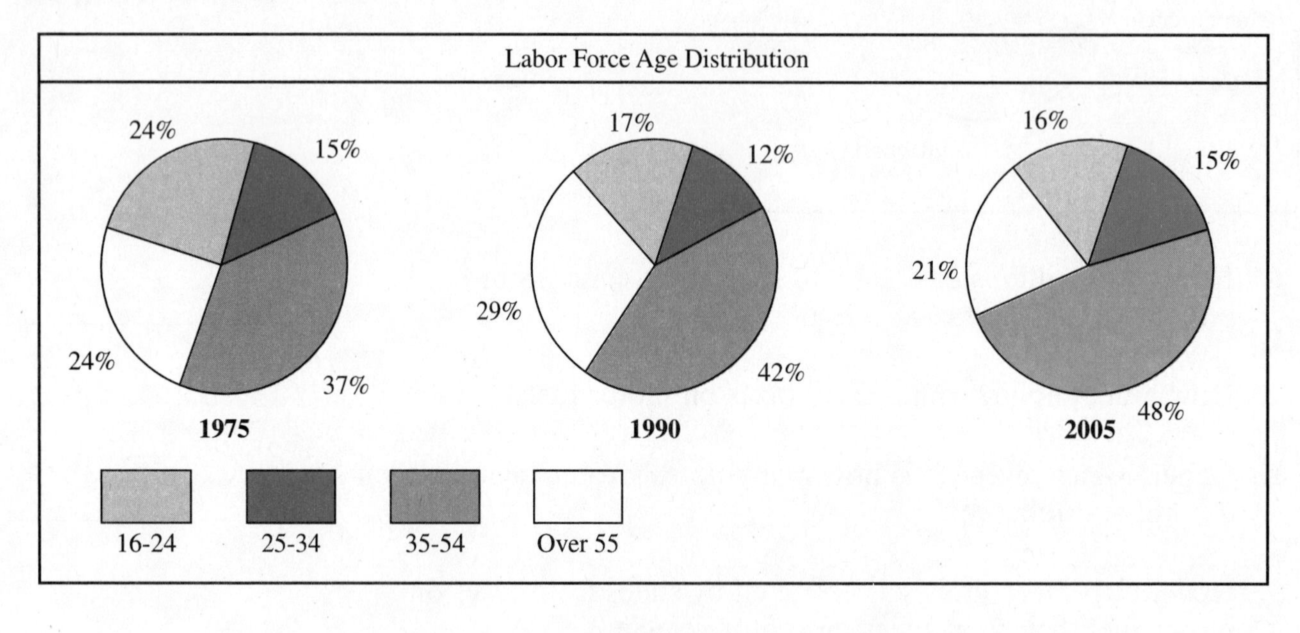

Example

Which age group is expected to increase between 1975 and 2005? The graphs show that the age group 35–54 is expected to increase from 37% in 1975 to 48% in 2005.

Practice

Use the circle graphs above to answer the following questions.

1. Did the percentage of people in the age group 16–24 increase or decrease from 1975 to 1990? ____________

2. Which age group had the least number of people in 1975 and in 1990? ____________

3. Which age group had a percentage decrease from 1975 to 1990 and is expected to return to its original percentage in 2005? ______

4. Which age group increased from 1975 to 1990 and is expected to decrease by 2005? ______

5. What is the difference in percent of people in the 16–24 age group between 1975 and 2005? ______

Problem Solving

Use the circle graphs on page 126 to solve the following problems.

6. If the total number of people in the labor force was 116 million in 1975, how many people were in the 25–34 age group? ______

7. If the total number of people in the labor force was 127 million in 1990, how many people were in the 25–34 age group? ______

8. Use your answers above to find the difference in the number of people in the 25–34 age group between 1975 and 1990. ______

LIFE SKILL

Planning a Budget

The Soos organize their monthly expenses.
Their take-home income is $3,600 per month.

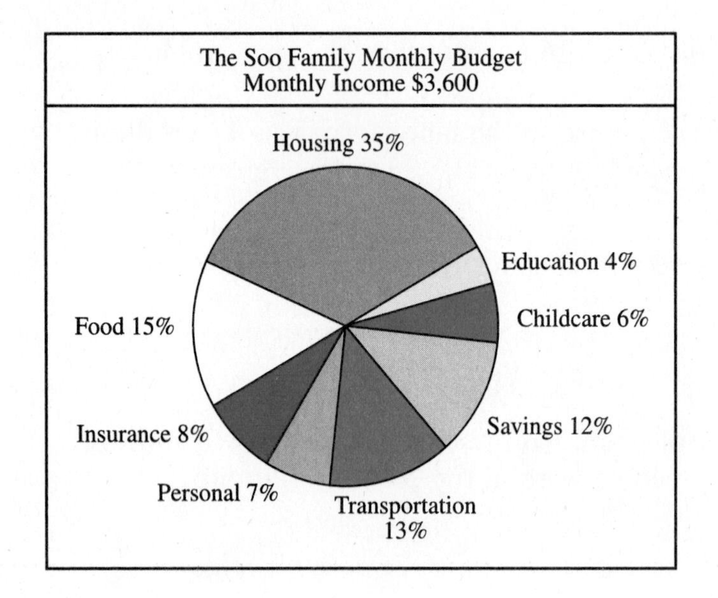

Use the circle graph to answer the following questions.

1. What percent of the Soo's income is set aside for savings? ______________

2. What is the largest item in their budget? ______________

3. What percent of their income do they spend on education and childcare? ______________

4. How much money do the Soos spend every month on transportation? ______________

5. How much money do they spend per month on insurance? ______________

6. How much money do they spend on food in one year? ______________

7. What part of their budget would include money spent on utilities? ______________

8. What other items could the Soos add to their budget? ______________

LESSON 41

Problem Solving—Using Circle Graphs

Circle graphs are useful problem-solving tools. When solving word problems, including those involving circle graphs, keep in mind the following steps:

Step 1 Read the problem and underline the key words. These words will generally relate to some mathematics reasoning computation.

Step 2 Make a plan to solve the problem. Ask yourself, Should I add, subtract, multiply, divide, round, or compare? You may have to do more than one of these operations for the same problem.

Step 3 Find the solution. Use your math knowledge to find your answer.

Step 4 Check the answer. Ask yourself, Is the answer reasonable? Did you find what you were asked for?

Example

Use the circle graph to solve the problem.

There are 9,844,000 male technicians. How many female technicians are there?

Step 1 Determine how many female technicians there are. The key words are **how many female technicians.**

Step 2 Since the problem gives the number of male technicians, the key words indicate which operation should occur—subtraction. Before you can subtract, you must find the total number of technicians.

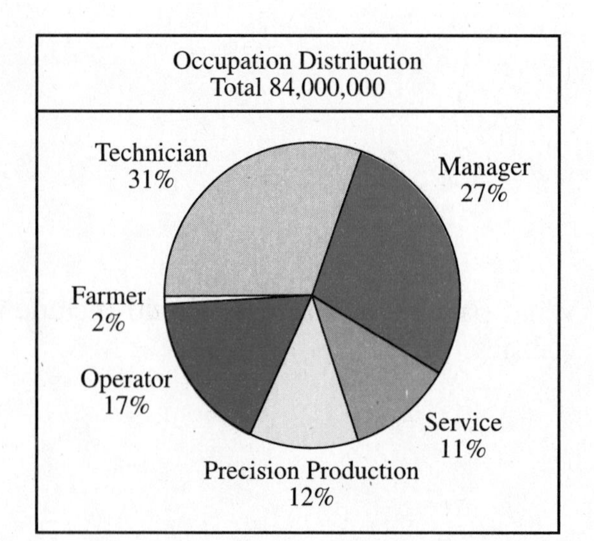

MATH HINT

To find the value of a percent in a circle graph, set up a proportion and solve for *n*.

Step 3 Find the solution.

$$\frac{n}{84{,}000{,}000} = \frac{31}{100}$$

Set up a proportion.

$$n = \frac{84{,}000{,}000 \times 31}{100}$$

Multiply the known diagonals and divide by the number that is left.

n = 26,040,000 technicians

total	26,040,000
male	− 9,844,000
female	16,196,000

Subtract to find the number of female technicians.

There are 16,196,000 female technicians.

Step 4 Check the answer. Does it make sense that there are 16,196,000 female technicians? Yes, the answer is reasonable.

Practice

Use the circle graph on page 130 to solve the following problems.

1. How many operators are there? __________

2. How many more operators are there than service employees? __________

3. If there are 11,165,000 female managers, how many male managers are there? __________

4. Precision production has how many times more workers than there are farmers? __________

5. In the service industry, 50% of the employees are men. How many women work in the service industry? __________

6. If one in every ten farmers is a woman, how many farmers are women? __________

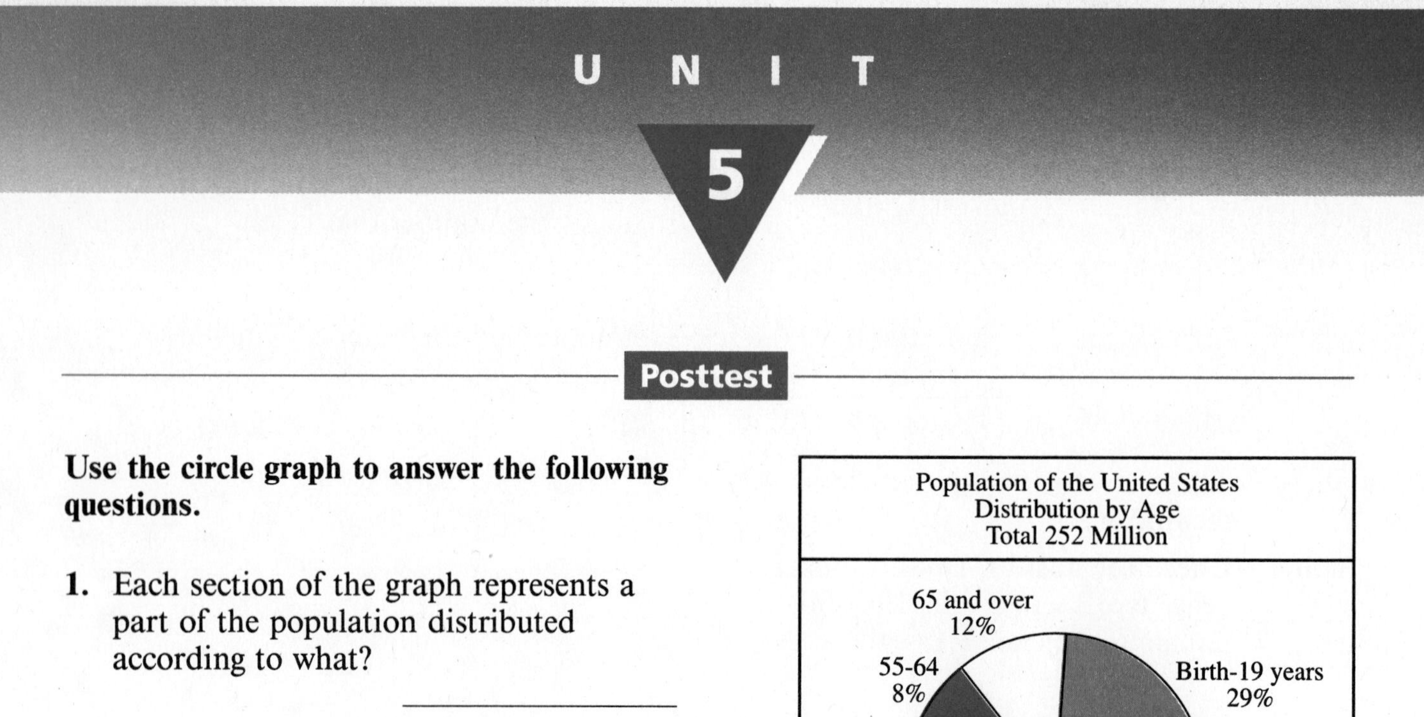

UNIT 5

Posttest

Use the circle graph to answer the following questions.

Population of the United States
Distribution by Age
Total 252 Million

65 and over 12%
55-64 8%
Birth-19 years 29%
45-54 10%
35-44 16%
20-34 25%

1. Each section of the graph represents a part of the population distributed according to what? ____________

2. What is the smallest age group? ____________

3. What percent of the population is between the ages of 45 and 54? ____________

4. What percent of the population is younger than 35? ____________

5. How many people are between the ages of 20 and 34? ____________

6. How many people are 65 or older? ____________

7. How many more people are in the 35–44 age group than in the 55–64 age group? ____________

8. How many people are older than 54? ____________

UNIT

Statistics

Pretest

Below is a list of driving test scores. Refer to the list to solve the following problems.

82	78	69	93	80
84	65	82	82	84
71	97	86	100	85
95	63	90	77	71

1. Make a frequency table of the test scores. The intervals are given.

Scores	Tally	Frequency
61–70		
71–80		
81–90		
91–100		

2. What is the mean score? ________

3. What is the median score? ________

4. What is the mode? ________

5. What is the range? ________

Problem Solving

Use the frequency table on page 133 to solve the following problems.

6. What is the frequency of scores in the 81–90 interval? ________________

7. How many people scored below 71? ________________

8. What percent of the people scored 91 or higher? ________________

9. Anyone who scored lower than 71 had to retake the test. What percent of the class had to retake the driving test? ________________

10. Andie's score was 81. How many people scored the same or higher than she? ________________

Introduction to Statistics

Statistics is the study of collecting, organizing, and analyzing data or pieces of information. You can collect data by observing or asking questions. Once the data is collected, you can organize it. One of the most common ways to organize data is to record the information in a table or chart. The chart at the right shows the test scores of five students.

Student	Score
Craig	83
Andie	91
Brad	82
George	75
Drew	98

Another way to organize data is to make a **frequency table.** A frequency table shows the number of values occurring in each interval. The frequency table at the right shows the number of hours worked by 20 employees in one week. It also shows a tally of the number of employees who worked certain hours. The tallies are added to make the frequency.

Number of Hours Worked	Tally	Frequency
0–9	||	2
10–19	|||	3
20–29	||||	4
30–39	~~||||~~ |||	8
40–44	|||	3

To make a frequency table, follow these steps:

Step 1 Choose the intervals for grouping the data and write them in the first column. The intervals should be equal widths.

Step 2 Enter a tally mark in the second column for each value occurring in an interval.

Step 3 Count the number of tally marks in each row and write the numbers in the third column.

MATH HINT

Five tallies are marked ~~||||~~ to make it easier to keep track of the count.

Examples

A. Organize the following data into a frequency table. The data represents high temperatures in degrees Fahrenheit for 20 cities.

96	105	95	98	100
97	99	96	92	101
91	95	89	104	106
102	93	100	94	86

MATH HINT

Rewrite the data in numerical order to make it easier to organize the data into a frequency table.

Step 1 Before you choose the intervals, rewrite the data in numerical order:

86, 89, 91, 92, 93, 94, 95, 95, 96, 96, 97, 98, 99, 100, 100, 101, 102, 104, 105, 106.

Choose the intervals 85–89, 90–94, 95–99, 100–104, and 105–109; and write them in the first column of the table.

Step 2 Enter a tally mark for each value occurring in the intervals.

Step 3 Count the number of tally marks for each interval and write the numbers in the third column.

Temperature (Step 1)	Tally (Step 2)	Frequency (Step 3)
85°–89°	II	2
90°–94°	IIII	4
95°–99°	~~IIII~~ II	7
100°–104°	~~IIII~~	5
105°–109°	II	2

B. What is the frequency of temperatures between 95° and 99°? The third row of the table indicates that the frequency of temperatures between 95° and 99° is 7.

Data can be analyzed using the **range.** The range is the difference between the greatest and the least values.

C. What is the range of the temperatures shown in Example A? The greatest temperature is 106°. The least temperature is 86°.

$106° - 86° = 20°$ Subtract to find the difference.
The range is 20°.

Practice

Make a frequency table for the following list of weights.

1. 201, 355, 472, 189, 294, 222, 500, 157, 120, 290, 438, 419, 499, 473, 402, 78

Weight in Pounds	Tally	Frequency
1–100		
101–200		
201–300		
301–400		
401–500		

Problem Solving

Solve the following problems.
Problems 2–7 are related.

2. In order to use the harbor club boats, everyone has to pass a water safety test. Below are the test scores of 25 people. Organize the scores into a frequency table.

38	29	35	36	40
27	35	35	33	31
35	29	38	38	40
27	34	35	31	34
35	31	27	30	31

Scores	Tally	Frequency
25–28		
29–32		
33–36		
37–40		

3. What is the frequency for the scores of 29–32? ______________

4. What is the range of the test scores? ______________

5. What is the frequency of the scores that occurred most often? ______________

6. If a score of 29 or more is passing, how many people passed? ______________

7. What percent of the people passed the test? ______________

LIFE SKILL

Taking a Survey

A **survey** is an example of collecting and organizing statistical data. A company is designing a new pair of walking shoes and is taking a survey at a shopping mall to learn what factor people consider most important when purchasing shoes: comfort, quality, price, or appearance.

Below is a frequency table displaying the results of the survey.

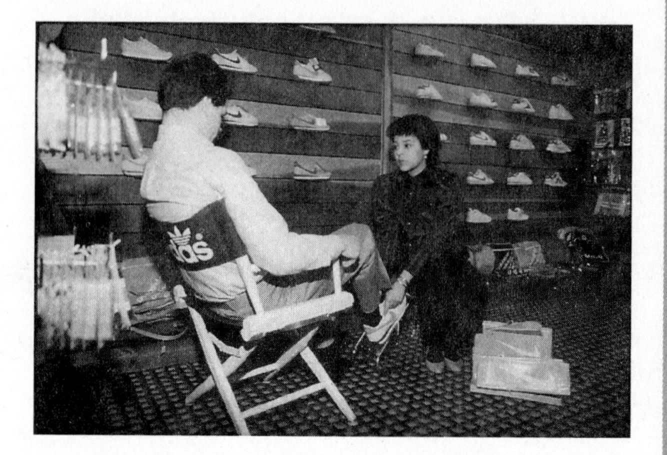

Factor	Tally	Number of People
Comfort	||||/ ||||/ ||||/ ||||/		21
Quality	||||/ ||||/ ||||	14
Price	||||/ ||||	9
Appearance	||||/ ||||/ ||||/		16

Use the table to answer the following questions.

1. How many people were surveyed? __________

2. Which category received the greatest number of responses? __________

3. How many people considered appearance the most important factor? __________

4. How many people considered quality the most important factor? __________

5. What percent of the people surveyed considered price the most important factor when purchasing new shoes? __________

6. How could the company use this information in advertising to sell more shoes? __________

LESSON 43

Histograms

A **histogram** is a bar graph that shows the frequency of certain values occurring. You can analyze data when it is organized into a histogram. To construct a histogram, follow these steps:

Step 1 List the intervals along the bottom of the graph.

Step 2 Write the scale along the left side of the graph.

Step 3 Fill in the boxes that represent the frequency of values occurring in each interval.

Examples

A. Students were asked to record the number of hours they read each week. The data was collected and organized into a frequency table such as the one shown below. Make a histogram of the data using the steps outlined above.

Hours	Tally	Frequency
0–2	~~||||~~ ~~||||~~ ||	12
3–5	~~||||~~ ~~||||~~ ~~||||~~ |||	18
6–8	~~||||~~ ~~||||~~ ~~||||~~ ~~||||~~ ~~||||~~ ~~||||~~ ||	32
9–11	~~||||~~ ~~||||~~ ~~||||~~ ~~||||~~ ||	22
12–14	~~||||~~ ~~||||~~ ||	12
15–17	||||	4

Step 1 List the intervals along the bottom of the graph.

Step 2 Write the scale along the left side. In this histogram, each horizontal line represents two students.

Step 3 Fill in the boxes that represent the frequency of values occurring in each interval. There are 12 students who read 0–2 hours a week. Fill in the boxes to the line marked 12. Continue to fill in the boxes for each interval.

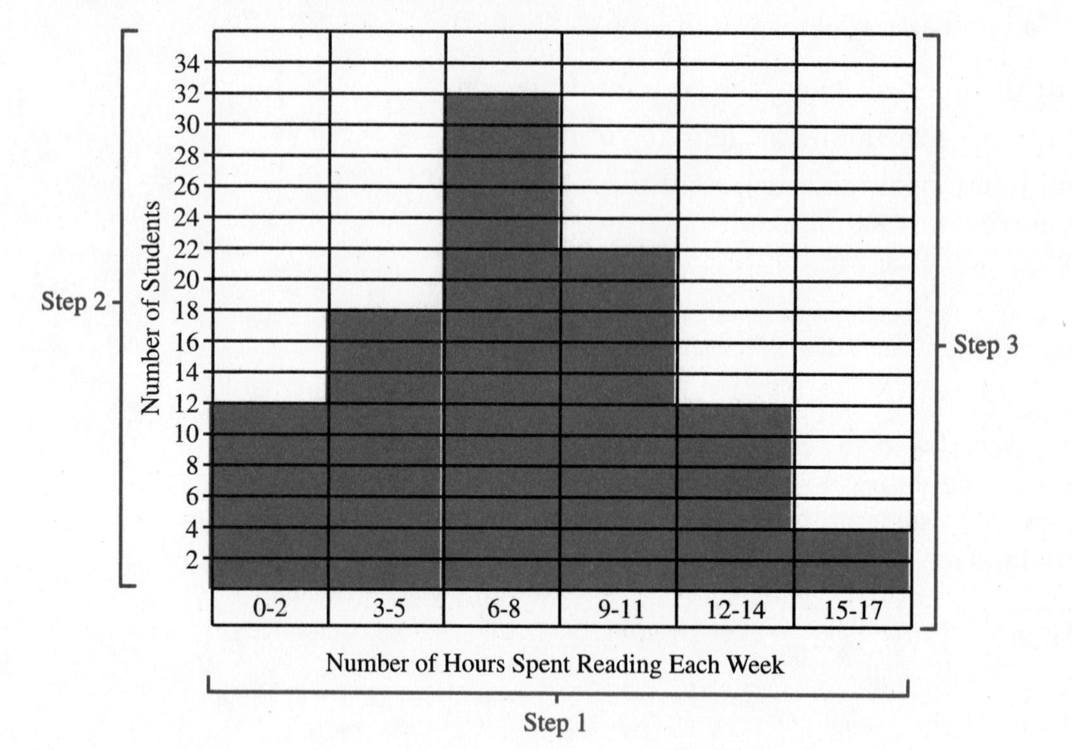

B. How many students read 12 or more hours each week? According to the histogram, 12 students read 12–14 hours and 4 students read 15–17 hours.

$12 + 4 = 16$ Add to find the total.

16 students read 12 or more hours each week.

Practice

Use the histogram in the example on page 140 to answer questions 1–9.

1. How many hours did the greatest number of students read? ______________
2. How many hours did the fewest number of students read? ______________
3. How many people were surveyed? ______________
4. Which two intervals had equal frequencies? ______________
5. How many students read 5 hours or less? ______________
6. What percent of the students read 8 hours or less? ______________
7. What percent of the students read 9 hours or more? ______________
8. If the survey was done in a grocery store instead of a school, do you think the answers would be the same or different? ______________
9. Why or why not?

____

Make a histogram of the data in the frequency table below and use it to answer questions 11–15.

10. The Lifetimes of Light Bulbs in Days

Days	Tally	Frequency
150–199	卌 III	8
200–249	卌 卌 III	13
250–299	卌 卌 IIII	14
300–349	卌 卌 卌 III	18
350–399	卌 卌 卌	15
400–449	卌 卌 II	12

11. How many light bulbs lasted between 250 and 299 days? ______

12. In what interval did the greatest number of light bulbs last? ______

13. How many light bulbs are represented in the graph? ______

14. What percent of the light bulbs lasted between 250 and 349 days? ______

15. What percent of the light bulbs lasted 350 days or more? ______

LESSON 44

Measures of Central Tendency

It is useful to have a number that represents a whole set of data. Numbers known as **measures of central tendency** are often used to represent data since they represent middle values, or the center, of the data.

The three most common measures of central tendency are the mean, the median, and the mode.

The **mean** is the average of the data.
The **median** is the middle value in a set of ordered data.
The **mode** is the number that occurs most often in a set of data.

Examples

The Rodriguez family did some comparison shopping before they bought a radio. They compared the price of the same brand at seven shops. Below is the list of prices from the lowest to the highest. Find the mean, median, and mode of the set of data.

$47, $48, $50, $52, $52, $53, $55

A. The **mean**, or average price, is the total of all the prices divided by the number of prices.

$$\frac{\$47 + \$48 + \$50 + \$52 + \$52 + \$53 + \$55}{7} = \frac{\$357}{7} = \$51$$

The mean does not have to be one of the numbers listed.

B. The **median** is the middle value of a list of values written in numeric order.

$47	$48	$50	$52	$52	$53	$55
1	2	3	4	5	6	7

MATH HINT

If the number of values in the list is an even number, then the average of the two middle numbers is the median.

In a list of 7 numbers, the middle number is fourth.
The median is $52.
The median does not have to be one of the numbers listed.

C. The mode is the value that appears the most often. $52 is the only price that occurs twice, so it is the mode. The mode is always one of the numbers given.

> **MATH HINT**
>
> If two values have the same high frequency, then each value is a mode.

The mean, or average, price is $51.
The median, or middle, price is $52.
The mode, or most common, price is $52.

47 48 49 50 51 52 53 54 55

Mean Median

Mode

The three measures of central tendency are marked on the number line at the right.

Practice

Look at the list of typing scores, expressed in words per minute (wpm), for currently employed data entry clerks in a company.

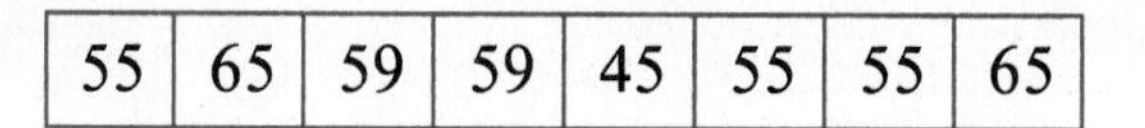

55	65	59	59	45	55	55	65

Answer the following questions.

1. Rewrite the scores in order from largest to smallest.

Refer to the scores in problem 1 to answer questions 2–9.

2. What is the lowest score? ________
3. What is the highest score? ________
4. What is the mean score? ________
5. What is the median score? ________
6. What is the mode? ________

Two new people were hired, and they both scored 65 wpm on the typing test.

7. What is the new mode? ________
8. What is the new median? ________
9. What is the new mean? ________

Below is a set of mileage rates for a new car in a wide variety of road conditions.

21	30	17	19	21
18	27	30	21	21
22	30	28	29	21
29	25	28	29	22
27	17	17	30	21

Use the table above to answer the following questions.

10. What is the mean? __________

11. What is the median? __________

12. What is the mode? __________

13. Mark the mean, median, and mode on the number line.

17 18 19 20 21 22 23 24 25 26 27 28 29 30

LESSON 45

Problem Solving—Statistics

Review the steps for problem solving:

Step 1 Read the problem and underline the key words. These words will generally relate to some mathematics reasoning computation.

Step 2 Make a plan to solve the problem. Ask yourself, Should I add, subtract, multiply, divide, round, or compare? You may have to do more than one of these operations for the same problem. You may also be able to estimate your answer.

Step 3 Find the solution.

Step 4 Check the answer. Ask yourself, Is this answer reasonable? Did you find what you were asked for?

Here are some key statistics words to remember:

frequency	mean
range	median
average	mode

Example

A manufacturing company is trying to decide whether it would be beneficial to build an exercise facility for its employees. The company officials want to know the average number of hours the employees exercise each week. By taking a poll, they find that 4 people exercise 5 hours per week, 6 people exercise 4 hours, 8 people exercise 3 hours, and 2 people exercise 1 hour. What is the average number of hours that each person exercises per week?

Step 1 Determine the average number of hours each person exercises each week. The key word is **average.**

Step 2 The key word indicates that you should find the mean.

Step 3 Find the solution. The mean is the total hours divided by the total number of people.

$$
\begin{aligned}
&4 \text{ people} \times 5 \text{ hours per week} = 20 \text{ hours} \\
&6 \text{ people} \times 4 \text{ hours per week} = 24 \text{ hours} \\
&8 \text{ people} \times 3 \text{ hours per week} = 24 \text{ hours} \\
&\underline{2 \text{ people}} \times 1 \text{ hour per week} = \underline{\ \ 2 \text{ hours}} \\
&20 \text{ people} \qquad\qquad\qquad\qquad\quad 70 \text{ hours}
\end{aligned}
$$

Find the total number of people and the total number of hours.

$$\frac{70 \text{ total hours}}{20 \text{ number of people}}$$

Divide.

$= 3.5$ hours

The average number of hours each person exercises per week is 3.5 hours.

Step 4 Check the answer. Does it make sense that the average number of hours each person exercises per week is 3.5 hours? Yes, the answer is reasonable.

Practice

Solve the following problems.

1. Rhonda's weekly chemistry quiz scores are given below:

20, 18, 18, 21, 19, 25, 23, 21, 17, 23

(1) What is her average quiz score? ____________

(2) What percent of the scores is 21 or higher? ____________

(3) What is the range of her scores? ____________

(4) If she scores a 25 on her next quiz, what will her new average be? ____________

2. A survey question asked people to rank the car they test drove from 1 to 7, with 1 being the lowest score and 7 being the highest. The results of the survey are below:

1, 1, 1, 2, 2, 2, 2, 2, 3, 3, 3, 3, 4, 4, 4, 5, 5, 5, 5, 6, 6

(1) What is the mean ranking? ____________

(2) How many people gave their test car a rank of 5? ____________

(3) What percent of the scores were greater than 3? ____________

(4) What ranking did the greatest number of people give to the cars they drove? ____________

3. The local car wash has 10 employees. Three people earn $5.20 per hour, two earn $4.80, four earn $5.50, and one earns $5.90.

(1) What is the hourly wage earned by the greatest number of people? ______________

(2) What is the average wage of the employees? ______________

(3) What is the range of the wages? ______________

(4) If two new managers are hired at $6.10 and $6.30 an hour, what will the new average wage be? ______________

UNIT

6

Posttest

Solve the following problems.

The Kendalls compared prices at many stores before buying a television. Below is a list of prices they found:

$430	$460	$469	$465	$456	$417
$456	$470	$499	$463	$435	$476

1. Make a frequency table of the prices. The intervals are given.

Prices	Tally	Frequency
401–420		
421–440		
441–460		
461–480		
481–500		

2. Make a histogram of the frequency table in problem 1.

5					
4					
3					
2					
1					
	______	______	______	______	______

Prices

3. What is the mean price? __________

4. What price is the mode? __________

5. What is the range of prices? __________

6. What is the median price? __________

Problem Solving

Solve the following problems.

7. The number of points scored by the basketball team in the past eight games were 84, 62, 71, 75, 65, 70, 67, and 81.

(1) What is the median? ______________

(2) What is the mean, to the nearest whole point? ______________

8. Ten students' scores on a 30-point algebra quiz are 28, 25, 19, 24, 30, 20, 21, 25, 26, and 16.

(1) What is the frequency of students who scored 25 or more? ______________

(2) What is the range of the quiz scores? ______________

(3) What is the mode? ______________

Problems 9 and 10 are related.

9. Melanie received the following tips each day waiting tables: Sunday, $45; Monday, $25; Tuesday, $27; Wednesday, $38; Thursday, $26; Friday, $50. What is the mean amount that she received? ______________

10. If Melanie gets $55 in tips on Saturday, what will the new mean amount be? ______________

UNIT

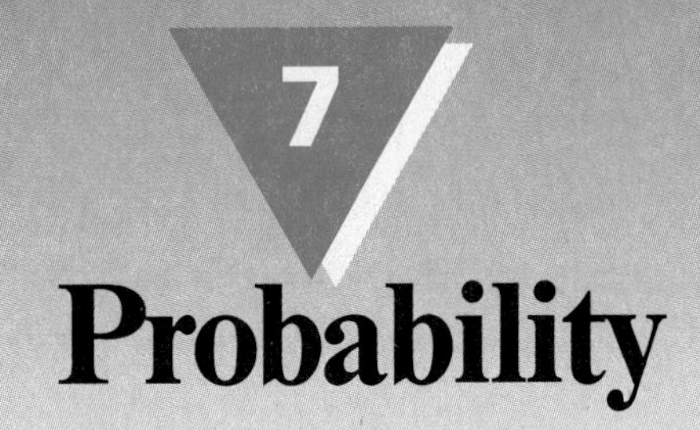

Probability

Pretest

Use the following information to solve problems 1–6. Write probabilities as fractions reduced to lowest terms.

There are 24 buttons in a drawer. The buttons are classified as follows:

6	red
12	blue
2	green
4	yellow

Assume that you reach into the drawer without looking and always replace any buttons you take out.

1. What is the probability of selecting a green button? ________

2. What is the probability of selecting a red button? ________

3. What is the probability of selecting a blue or a green button? ________

4. What is the probability of selecting a yellow and a red button? ________

5. What is the probability of selecting a button that is not blue? ________

6. What is the probability of selecting 2 yellow buttons? ________

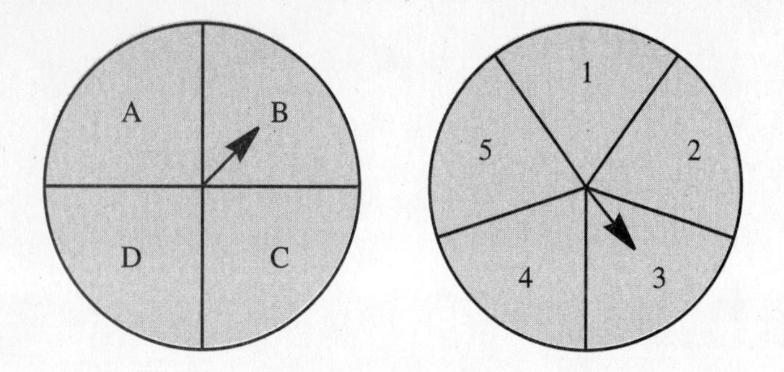

Use the following information to solve problems 7–13.

The two spinners above are each divided into parts. The first one is labeled A, B, C, D. The second one is labeled 1, 2, 3, 4, 5.

7. In a game, a player must spin each spinner. Make a table of all the different outcomes that are possible when each spinner is spun separately. (Hint: A1 is an outcome.)

8. How many possible outcomes are there? __________

9. What is the probability of spinning a 5 and an A? __________

10. What is the probability of spinning an A and an odd number? __________

11. What is the probability of spinning an A and an even number? __________

12. What is the probability of spinning B or C and an even number? __________

13. What is the probability of spinning a D and a number greater than 1? __________

Problem Solving

Solve the following problems.

14. A conservation group captured and tagged 30 wild geese. Two weeks later they captured 40 wild geese. Three of those geese were tagged. What is the estimated population of geese in the area? ________________

15. A store manager asked a random number of people who came in to buy soda if they preferred soda made with sugar or soda made with a noncalorie sweetener. Of the 100 people he asked, 40 preferred the soda made with sugar and 60 preferred the noncalorie sweetener. One thousand people come into his store each week to buy soda. What is the estimated number of people who will buy soda made with sugar? ________________

LESSON 46

Introduction to Probability

Probability is the chance that an event will happen.

You may have heard someone say:

- There is a 100% chance of rain.
- The probability of winning the raffle is 1 in 10,000.

That means the same as:

- It will rain.
- 10,000 raffle tickets have been sold, and I have one of them. There is very little chance that I will win.

The probability of an event is the ratio of the number of ways an event can occur to the total number of possible outcomes.

$$\text{Probability} = \frac{\text{number of ways an event can occur}}{\text{number of possible outcomes}}$$

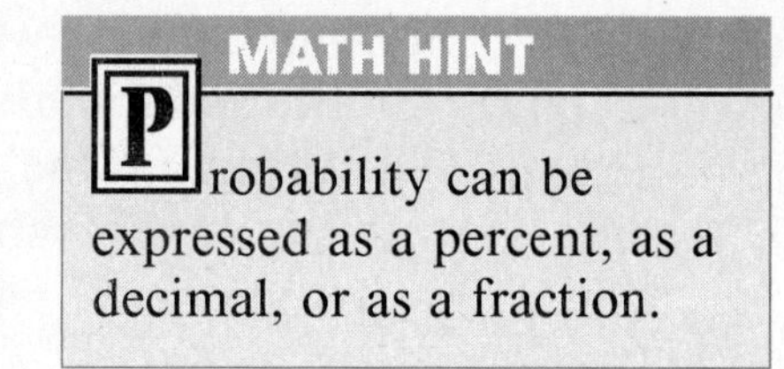
MATH HINT

Probability can be expressed as a percent, as a decimal, or as a fraction.

Examples

A. What is the probability of flipping a coin and having it land heads up?

There is 1 head on a coin, so the number of ways that heads can occur is 1.

There are two possible outcomes when flipping a coin: heads and tails.

So, the number of possible outcomes is 2.

$$\text{Probability} = \frac{\text{number of ways an event can occur}}{\text{number of possible outcomes}} = \frac{1}{2}$$

The probability of flipping a coin and having it land heads up is one out of two, or $\frac{1}{2}$.

B. If a die is rolled, what is the probability of rolling a number less than 3?
There are 2 numbers on a die less than 3. So the number of ways for the event to occur is 2. There are 6 sides of a die, so the number of possible outcomes is 6.

$$\text{probability} = \frac{\text{number of ways an event can occur}}{\text{number of possible outcomes}} = \frac{2}{6}$$

$$= \frac{1}{3} \quad \text{Reduce.}$$

MATH HINT

Probabilities written as fractions should be reduced.

The probability of rolling a number less than 3 is 1 out of 3, or $\frac{1}{3}$.

Probability always has a value between 0 and 1.

A probability of zero means there is no chance of an event happening.

A probability of one means that there is every chance that an event will happen.

Examples

C. What is the probability of rolling a 7 with one die? Since this event is impossible, the probability is 0.

D. What is the probability that either a head or a tail will show when a coin is flipped? Since this event is certain to happen, the probability is 1.

Practice

Answer the following questions. Write the probabilities as fractions reduced to lowest terms.

Questions 1–3 are related.

1. How many sides of a coin has a tail? __________

2. What is the total number of sides on a coin? __________

3. What is the probability of flipping a coin and getting a tail? __________

Questions 4–7 are related.

4. How many possible choices are there for each question on a true and false test? __________

5. What is the probability of true being the correct choice for a question? __________

6. What is the probability of false being the correct choice for a question? __________

7. What is the probability of getting a question right if you leave it blank? __________

Questions 8–14 are related. One die is rolled.
(Hint: A die has six sides.)

8. What is the probability of rolling a 5? __________

9. What is the probability of rolling a 3? __________

10. What is the probability of rolling a number greater than 3? __________

11. What is the probability of rolling an even number? __________

12. What is the probability of rolling an odd number? __________

13. What is the probability of rolling a 7? __________

14. What is the probability of rolling a number greater than 0 and less than 7? __________

LESSON 47

Adding Probabilities

You can find the probability of events that cannot occur at the same time by adding their individual probabilities.

Examples

A. The spinner below is equally divided into five parts: A, B, C, D, and E.

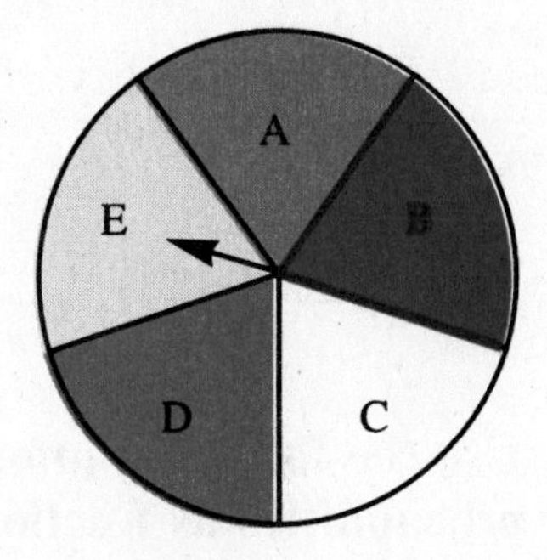

What is the probability of spinning an A or a B? Since spinning an A and a B cannot occur at the same time, you can add to find the probability.

There is one A, so there is one way to spin A.
There are 5 parts, so there are 5 possible outcomes.
The probability of spinning an A is $\frac{1}{5}$.
The probability of spinning a B is also $\frac{1}{5}$.

$\frac{1}{5} + \frac{1}{5} = \frac{2}{5}$ Add the probabilities.

The probability of spinning an A or a B is $\frac{2}{5}$.

B. What is the probability of **not** spinning a D? There are 4 parts that are not D. The probability of spinning each of them is $\frac{1}{5}$.

$\frac{1}{5} + \frac{1}{5} + \frac{1}{5} + \frac{1}{5} = \frac{4}{5}$ Add to find the probability of spinning a letter other than D.

The probability of **not** spinning a D is $\frac{4}{5}$.

C. What is the probability of spinning an A, B, C, D, or E?
The probability of spinning an A is $\frac{1}{5}$.
The probability of spinning a B is $\frac{1}{5}$.
The probability of spinning a C is $\frac{1}{5}$.
The probability of spinning a D is $\frac{1}{5}$.
The probability of spinning an E is $\frac{1}{5}$.

$\frac{1}{5} + \frac{1}{5} + \frac{1}{5} + \frac{1}{5} + \frac{1}{5} = \frac{5}{5} = 1$ The probability of spinning an A, B, C, D, or E is 1.

MATH HINT

The probabilities of all the outcomes added together will equal 1.

D. There are 4 red pens, 5 green pens, 1 blue pen, and 6 black pens all mixed up in the bottom of a drawer. If you reach in and take a pen without looking, what is the probability of drawing a red or green pen?

There are 4 ways to draw a red pen.
There are 16 pens in all.
The probability of drawing a red pen is $\frac{4}{16}$.

There are 5 green pens.
The probability of drawing a green pen is $\frac{5}{16}$.

$\frac{4}{16} + \frac{5}{16} = \frac{9}{16}$ Add the probabilities.

The probability of drawing a red or green pen is $\frac{9}{16}$.

Practice

Use the following information to solve problems 1–5. Write the probabilities as fractions in reduced form.

The spinner to the right is equally divided into 9 parts. Each part has a number: 1, 2, 3, 4, 5, 6, 7, 8, 9. The pointer is equally likely to land on each segment. What are the probabilities of the following?

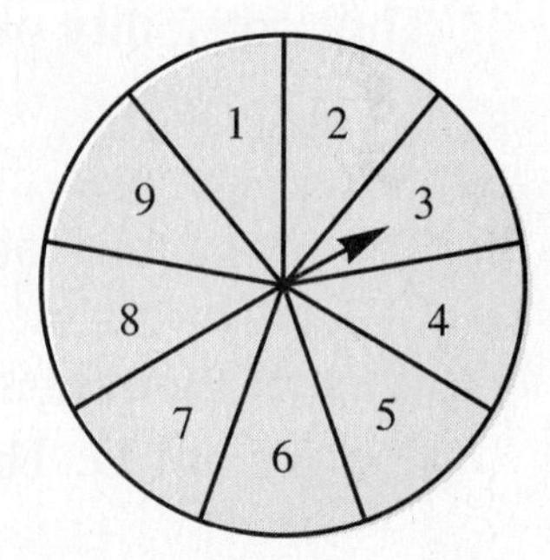

1. The spinner landing on the number 4? ________

2. The spinner landing on the numbers 5 or 8? ________

3. The spinner landing on an even number? ________

4. The spinner not landing on a number divisible by 3? ________

5. The spinner not landing on an odd number? ________

Use the following information to solve problems 6–10. Write the probabilities in reduced form.

The spinner to the right is equally divided into 6 parts. Each part has a letter: L, E, T, T, E, R. The pointer is equally likely to land on each segment. What are the probabilities of the following?

6. The spinner landing on the letter E? __________

7. The spinner landing on the letters E or T? __________

8. The spinner landing on the letters L or R? __________

9. The spinner not landing on T? __________

10. The spinner not landing on L or R? __________

Use the following information to solve problems 11–16. Write the probabilities in reduced form.

There are twelve marbles in a box. They are classified as follows:

3	black
4	red
2	white
3	yellow

Assume you are blindfolded and draw a marble from the box.

11. What is the probability of drawing a black marble? __________

12. What are the chances of drawing a white marble? __________

13. What is the probability of drawing either a black or a yellow marble? __________

14. What is the probability of drawing a red or a white marble? __________

15. What are the chances of drawing a green marble? __________

16. What is the probability of pulling out a marble? __________

LESSON 48

Random Samples

Probability is used to make predictions. By taking a small sample of a large group, probability can be used to predict how the large group will behave.

In order to use probability with a small sample, each member of the small sample must be selected randomly. **Random** means that no plan or system will affect the outcome of an event. So, each member will have an equal chance of being selected for the sample.

If the sample is random, then it is likely that the small sample will represent the large group. Then the probability of something occurring in the sample can be used as a proportion to predict what would happen in the large group.

Examples

In a randomly selected group of people in a grocery store between 4:00 p.m. and 8:00 p.m., 5 people said they were there to buy just ready-to-eat food for that night's dinner, 7 were there to buy groceries for a week, and 8 were there to pick up just a few necessities.

A. What is the probability that a person who is shopping between 4:00 p.m. and 8:00 p.m. is there to pick up a few necessities?

8 of the 20 people were there to pick up necessities.
The probability is $\frac{8}{20}$, or $\frac{2}{5}$.

B. Out of the next 1,000 people who go to a grocery store between 4:00 p.m. and 8:00 p.m., how many are likely to buy prepared foods for that night's dinner?

The random sample indicated that the probability was 5 in 20, or 1 in 4.

$\frac{1}{4} = \frac{n}{1{,}000}$ Set up a proportion.

$\frac{1 \times 1{,}000}{4} = 250$ Multiply the known diagonals and divide by the number that is left.

Out of 1,000 people, 250 of them would be likely to buy prepared food.

Practice

Write the probabilities in fraction form. Reduce to lowest terms.

In a random survey, 240 people indicated their most important reasons for selecting a car. The results are shown in the chart below.

Reason	Number of people
Price	80
Color	60
Performance	40
Model	20
Friend's recommendation	40
Total	240

1. What is the probability that a customer's main reason for purchasing a car is the price? __________

2. What is the probability that a customer's main reason for purchasing a car is the model? __________

3. What is the probability that a customer would base the selection on a friend's recommendation? __________

4. If 6,000 people are expected to buy new cars in the next year, how many people will choose their car based on color? __________

5. If 6,000 people are expected to buy new cars in the next year, how many people will choose their car based on performance? __________

6. If 4,800 people buy a new car, how many will buy that car based on price? __________

7. If you were a car dealer and wanted to put an advertisement in the newspaper, what would you emphasize in your ad? __________

LIFE SKILL

Using Random Samples and Probability to Predict Population

There are many practical uses of probability in the real world. One example is calculating probability to estimate the number of animals in a park.

Sylvia is a game warden at National Park. To estimate the number of fish in the lake, she captured a random sample of 150 fish, marked them, and then released them. Three days later she captured 300 fish and found that 15 of them were marked.

The probability of catching a marked fish in a sample was $\frac{15}{300}$. Sylvia used that probability in a proportion to predict the total population.

Probability	First sample	Second sample
Sample marked	150	15
Total	n	300

$\frac{150}{n} = \frac{15}{300}$ Set up a proportion.

$\frac{\overset{10}{\cancel{150}} \times 300}{\underset{1}{\cancel{15}}} = 3{,}000$ Multiply the diagonals and divide by the number that is left.

Sylvia estimated that there were 3,000 fish in the lake.

Solve the following problems using probabilities.

1. The game warden captured 50 birds that she tagged and released. Two days later she captured 100 birds, and 3 of those were tagged. How many birds can she estimate are in the forest? __________

2. Winter is coming, and Sylvia wants to know how many deer are in the forest. She captured, tagged, and released 20 deer. Three days later she captured 30 deer, 3 of which were tagged. How many deer can she estimate are in the forest? __________

3. The conservation group caught, tagged, and released 40 wild rabbits. Five days later they captured 50 rabbits and only one was tagged. What is the estimated number of rabbits in the wild? ________

Problems 4–6 are related.

4. The conservation group captured and tagged 100 frogs in a wetland area. The following week they caught 50 frogs, and 5 were tagged. What was the population of frogs in that wetland? ________

5. Two years later the conservation group captured 100 frogs and tagged them. A few days later they captured 50 frogs, and 4 were tagged. What was the new population of frogs? ________

6. Was the population of frogs increasing or decreasing? ________

LESSON 49

Bias

Probability assumes that the actions taking place are random. If there is something that affects the outcome, the event is **biased.** A biased event has a greater or less chance of happening than a random event.

Example

Suppose a coin has two heads. Then the events occurring from flipping the coin would be biased.

What is the probability of tossing a two-headed coin and getting a tail?

Since there is no tail, the probability is 0.

What is the probability of tossing a two-headed coin and getting a head?

Since there are two heads, the probability is 1.

Practice

Solve the following problems.

1. Ten cards are in a box. Some cards are larger than others. Is drawing a card out of the box a random or biased event? ______________

2. A die is weighted so it only rolls a 6. Are the probabilities of rolling each number on the die equal? ______________

Is rolling a number on this die a random or biased event? ______________

3. Fifty marbles are in a sack. If the person drawing them out is blindfolded, is the event of drawing a marble random or biased? ______________

4. A company that makes novelty sweatshirts wants to know whether people would buy their sweatshirts. If the company interviews 100 people at a shopping mall at different times throughout the day and on different days of the week, are the responses random or biased? ______________

5. A cultural community center wants to determine whether people would be interested in attending a free concert of classical music in the park. The center calls 200 people from all areas of the community to ask them if they would be interested. Is this a random sample or a biased sample of people?

____

If the center took a survey of music students from the university coming out of class, would this sample be random or biased?

____

6. A game has a spinner that is divided into four sections: red, blue, green, and yellow. The red section makes up one-half of the entire spinner. Is the probability of spinning each color equally likely? ______________

7. A company wants to know how many people are interested in buying a home exercise machine, so they interview people walking out of an athletic club. Is the event that one of these people is interested random or biased? ______________

LESSON 50

Multiplying Probabilities

Two events are **independent** if the outcome of one event does not affect the outcome of the other event. You can find the probability of two or more independent events occurring at the same time by multiplying the probabilities of each event.

Examples

A. In a game, a player must roll a die and spin a spinner like the one shown at the right. What is the probability of rolling a 4 and getting an A on the spinner?

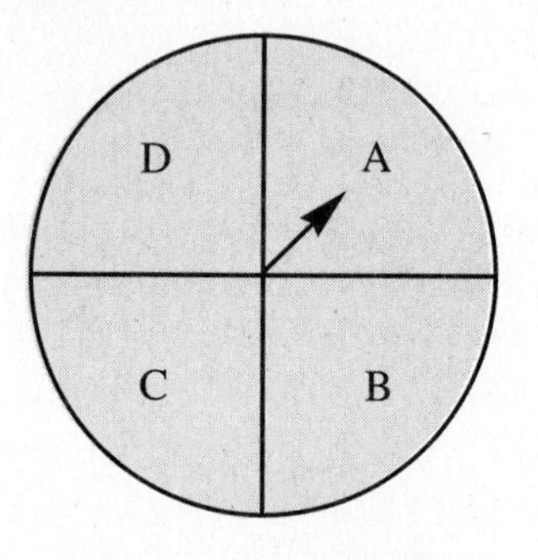

Each of these outcomes is independent since rolling a die does not affect spinning a spinner. Find the probability of each outcome. There is 1 chance in 6 of rolling a 4. The probability is $\frac{1}{6}$. There is 1 chance in 4 of spinning an A. The probability is $\frac{1}{4}$.

$\frac{1}{6} \times \frac{1}{4} = \frac{1}{24}$ Multiply.

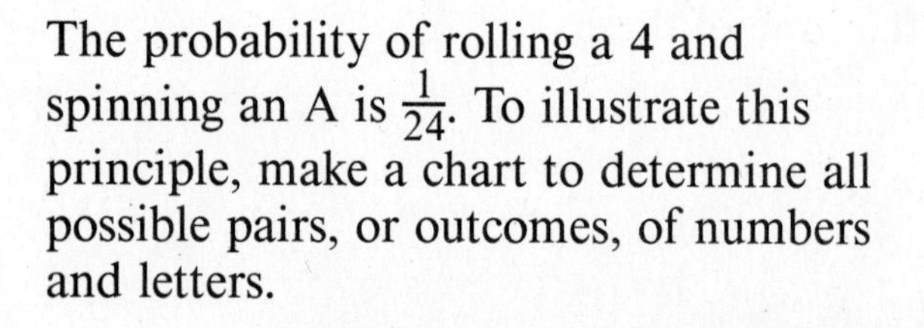

The probability of rolling a 4 and spinning an A is $\frac{1}{24}$. To illustrate this principle, make a chart to determine all possible pairs, or outcomes, of numbers and letters.

Die	Spinner	Die	Spinner	Die	Spinner	Die	Spinner
1	A	1	B	1	C	1	D
2	A	2	B	2	C	2	D
3	A	3	B	3	C	3	D
4	A	4	B	4	C	4	D
5	A	5	B	5	C	5	D
6	A	6	B	6	C	6	D

There is one combination of 4 and A on the chart.
That means there is 1 chance in 24 of spinning an A and rolling a 4, or a probability of $\frac{1}{24}$.

The figure below would be the same as a spinner divided into 24 parts. Each segment is labeled with one of the possible outcomes.

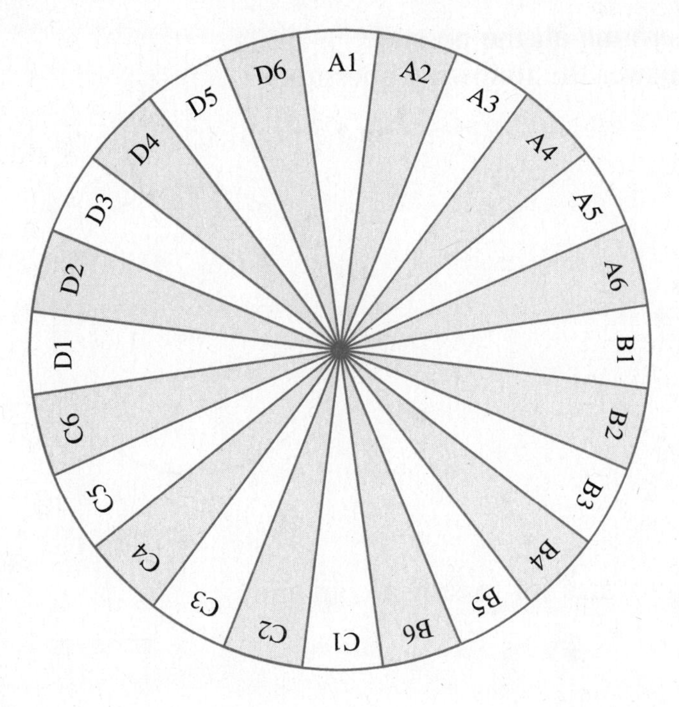

B. What is the probability of rolling an even number and spinning the letter B?

Look at the spinner above.
There are 3 even numbers with the letter B.
The probability is 3 in 24, or $\frac{3}{24} = \frac{1}{8}$.

You can also find the probability by multiplying.
The probability of rolling an even number is $\frac{3}{6}$, or $\frac{1}{2}$.
The probability of spinning a B is $\frac{1}{4}$.

$\frac{1}{2} \times \frac{1}{4} = \frac{1}{8}$ Multiply.

The probability is $\frac{1}{8}$.

Practice

Solve the following problems.

1. A game requires you to roll a die and toss a coin.

(1) Fill in the table below to determine all the possible outcomes. Use the table to answer the following questions.

Dice	Coin	Dice	Coin

Represent the problem in the circle below.

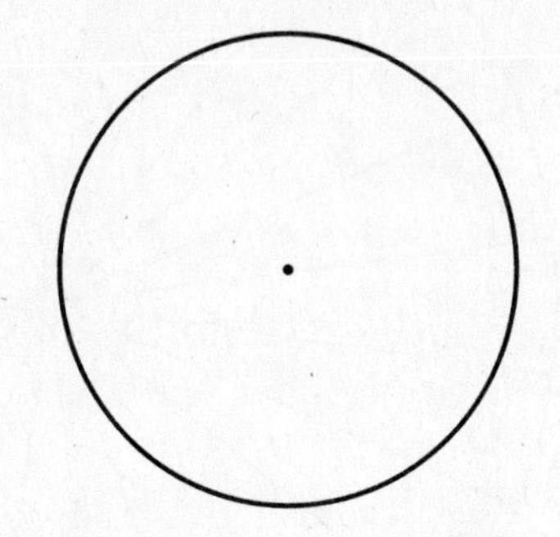

(2) What are the total possible outcomes for tossing a coin and rolling a die? ____________

(3) What is the probability of rolling a 6 and getting a head? ____________

(4) What is the probability of rolling either a 3 or a 4 and getting a tail? ____________

(5) What is the probability of rolling an even number and getting a head? ____________

(6) What is the probability of rolling a number greater than 3 and getting a tail? ____________

2. A game requires you to roll a die and spin a spinner like the one shown at the right. Use multiplication to find the following probabilities.

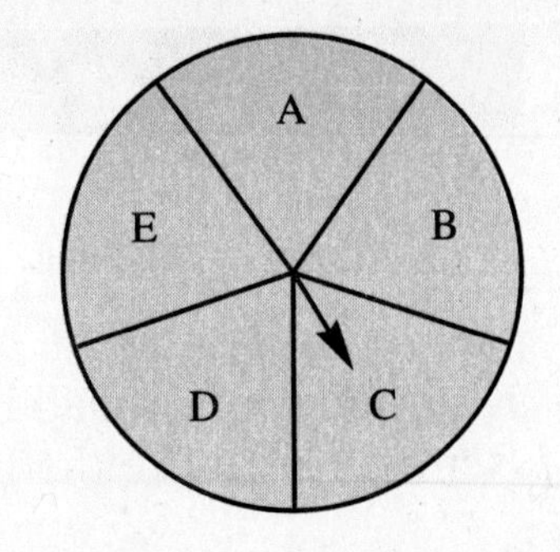

(1) What is the probability of rolling a 5 and spinning a C on one turn? ______________

(2) What is the probability of rolling either a 4 or a 5 and spinning an A? ______________

(3) What is the probability of rolling an odd number and spinning a D? ______________

(4) What is the probability of rolling a number greater than 4 and spinning either a B or a D? ______________

(5) What is the probability of not rolling a 5 and not spinning an E on one turn? ______________

LESSON 51

Problem Solving—Probability

Knowledge of probability is useful in solving word problems. Remember to use the following steps to solve word problems:

Step 1 Read the problem and underline the key words. These words will generally relate to some mathematics reasoning computation.

Step 2 Make a plan to solve the problem. Ask yourself, Should I add, subtract, multiply, divide, round, or compare? You may have to do more than one of these operations for the same problem.

Step 3 Find the solution. Use your math knowledge to find your answer.

Step 4 Check the answer. Ask yourself, Is the answer reasonable? Did you find what you were asked for?

When solving probability problems, remember the following:

1. The word **and** indicates that you should multiply probabilities.
2. The word **or** indicates that you should add probabilities.

Example

A die is rolled and a coin is tossed. What is the probability of getting a tail and rolling a 6?

Step 1 Determine the probability of getting a tail and rolling a 6. The key words are **probability** and **and.**

Step 2 The key words indicate which operation should occur—multiplication.

Before you can multiply, you must find the probability of each event occurring.
The probability of getting a tail is $\frac{1}{2}$.
The probability of rolling a 6 is $\frac{1}{6}$.

Step 3 Find the solution.

$\frac{1}{2} \times \frac{1}{6} = \frac{1}{12}$ Multiply the probabilities of each event occurring to find the probability of both events occurring at the same time.

Step 4 Check the answer. Does it make sense that the probability of getting a tail and rolling a 6 is $\frac{1}{12}$? Yes, the answer is reasonable.

Practice

Solve the following problems. Write probabilities as fractions in lowest terms.

1. The probability that Joyce will pass Spanish is $\frac{7}{8}$. The probability that she will pass geometry is $\frac{4}{5}$. What is the probability that she will pass Spanish and geometry? __________

2. A bag contains 3 black, 4 red, 2 white, and 3 yellow marbles. What is the probability of drawing a red or a white marble? __________

3. What is the number of possible outcomes for tossing two dice? __________

4. In one particular school, 20 children were randomly selected and asked if they had a certain video game at home. Of those children selected, 15 said yes, 3 said no, and 2 weren't sure. What is the probability that a child in the school will have the video game? __________

UNIT

7

Posttest

Use the following information to solve problems 1–5. Write probabilities as fractions reduced to lowest terms.

There are 36 socks in a drawer. They are classified as follows:

18	red
6	blue
2	black
10	white

Assume that you reach into the drawer without looking and always replace any socks you take out.

1. What is the probability of selecting a red sock? __________

2. What is the probability of selecting a blue or a black sock? __________

3. What is the probability of selecting two white socks? __________

4. What is the probability of selecting two black socks? __________

5. What is the probability of selecting a red sock or a white sock? __________

Use the following information to solve problems 6–10. Write probabilities as fractions reduced to lowest terms.

The two spinners at the right are each divided into parts. The first one is labeled 1, 2, 3, 4. The second one is labeled A, B, C, D, E.

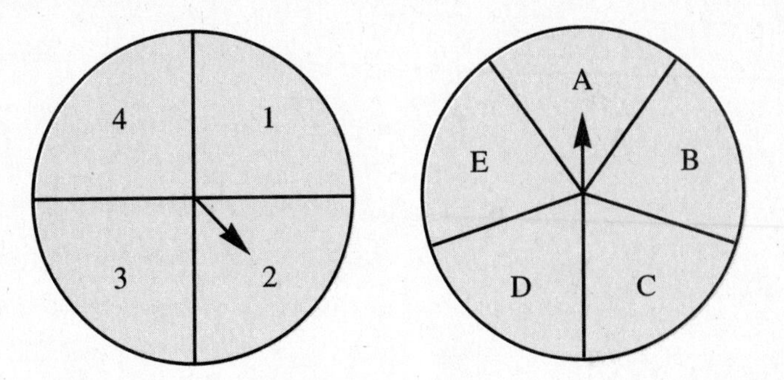

6. In a game, a player must spin each spinner. Make a table of all the different outcomes that are possible when each spinner is spun separately. (Hint: A1 is an outcome.)

7. What is the probability of spinning a 3 and an A? __________

8. What is the probability of spinning a 2 and a D? __________

9. What is the probability of spinning an A and an even number? __________

10. What is the probability of spinning B or C and an even number? __________

Problem Solving

11. A conservation group captured, tagged, and released 60 salmon. Two weeks later they captured 40 salmon. Eight of those salmon were tagged. What is the estimated population of salmon in the lake? __________

12. A store manager asked a random number of people who came in to buy bakery goods if they preferred bread made in the bakery or prepackaged bread. Of the 100 people he asked, 20 preferred bread made in the bakery and 80 preferred prepackaged bread. Five hundred people come into his store each week to buy bread. What is the estimated number of people who will buy bread made in the bakery? __________

ANSWERS

Unit 1 Pretest/pages 1-2

1. 14.921
2. 93.556
3. 202.852
4. 2,149
5. $9\frac{1}{10}$
6. $14\frac{11}{12}$
7. $1\frac{4}{7}$
8. $1\frac{3}{4}$
9. .2, 20%
10. $\frac{3}{50}$, 6%
11. $\frac{27}{50}$, .54
12. $\frac{13}{10}$, 130%
13. .375, 37.5%
14. $\frac{1}{500}$, .002
15. <
16. >
17. <
18. =
19. <
20. >
21. $1.65
22. $90,000
23. 900 people
24. 25%
25. 126 rooms

Lesson 1/pages 3-4

1. 215.79
2. 5.63
3. .99
4. 518.68
5. 13.48
6. 458.24
7. 13.779
8. 3.84 hours
9. 1.5 inches
10. $8.18
11. 1.6 hours
12. 2.6 hours

Lesson 2/pages 5-6

1. 25.83
2. 136.5
3. 554.692
4. 1.435
5. .2596
6. 210.951
7. 1,294.8
8. 430
9. $31
10. $49.95
11. $22.25

Lesson 3/pages 7-8

1. 620
2. 208
3. 42
4. 9,012
5. 1,700
6. .0095
7. .11 miles
8. $0.17
9. 19 salads

Lesson 4/pages 9-10

1. $\frac{2}{9}$
2. $\frac{1}{14}$
3. $12\frac{2}{7}$
4. $5\frac{3}{5}$
5. $18\frac{1}{3}$
6. $24\frac{1}{5}$
7. $\frac{10}{15}$
8. $\frac{16}{28}$
9. $\frac{21}{56}$
10. $\frac{23}{92}$
11. 28
12. 6
13. 55
14. 12
15. 15
16. 18
17. 18
18. 168
19. $\frac{1}{13}$
20. $\frac{2}{3}$
21. $\frac{18}{24}$, $\frac{16}{24}$, $\frac{3}{24}$

Lesson 5/pages 11-12

1. $\frac{1}{20}$
2. $\frac{5}{9}$
3. $\frac{3}{14}$
4. $31\frac{1}{2}$
5. 2
6. $\frac{1}{9}$
7. 5
8. $2\frac{2}{9}$
9. $2\frac{1}{7}$
10. $32\frac{1}{2}$
11. 7 yards
12. $12\frac{1}{2}$ feet
13. 54 people

Lesson 6/pages 13-14

1. $\frac{3}{5}$
2. 1
3. $13\frac{3}{4}$
4. $\frac{1}{2}$
5. $31\frac{1}{5}$
6. $\frac{1}{8}$
7. 59
8. $\frac{25}{42}$
9. $2\frac{2}{3}$
10. 10
11. $\frac{1}{27}$
12. $4\frac{2}{3}$
13. $\frac{3}{32}$ miles
14. $93\frac{3}{4}$ dispensers
15. 80 pieces

Lesson 7/pages 15-16

1. $12\frac{5}{11}$
2. $187\frac{1}{2}$
3. $9\frac{1}{2}$
4. $123\frac{21}{40}$
5. $6\frac{5}{9}$
6. $68\frac{5}{6}$
7. $6\frac{1}{4}$
8. $11\frac{17}{24}$
9. $37\frac{1}{2}$ feet

10. $23\frac{3}{8}$ miles
11. $12\frac{1}{2}$ feet
12. $12\frac{3}{4}$ miles
13. $41\frac{7}{8}$ feet

Lesson 8/pages 17-18

1. .25
2. .06
3. .003
4. .14
5. 1.3
6. .015
7. .0007
8. .09
9. $\frac{1}{5}$
10. $\frac{17}{100}$
11. $\frac{1}{25}$
12. $\frac{11}{200}$
13. $\frac{24}{25}$
14. $\frac{2}{3}$
15. $\frac{1}{400}$
16. $2\frac{2}{5}$
17. 60%
18. 40%
19. 1%
20. 62.5%
21. 150%
22. 500%
23. 60%
24. .9%
25. 17%
26. $\frac{3}{25}$
27. .05
28. 4%

Lesson 9/pages 19-20

1. >
2. <
3. >
4. <
5. <
6. >
7. =
8. <
9. <
10. =
11. <
12. =
13. 70%, $\frac{5}{8}$, 0.61
14. $1\frac{1}{2}$, 0.6, 1.2%
15. $\frac{4}{10}$, 0.045, 4%
16. $\frac{3}{5}$, 0.01, 0.6%
17. 1.3, 33%, $\frac{3}{10}$
18. 1.8, 150%, $1\frac{1}{5}$
19. 5%
20. Matthew
21. February
22. 78%, 0.75, $\frac{36}{50}$

Lesson 10/pages 21-22

1. 1:10, $\frac{1}{10}$
2. 3:1, $\frac{3}{1}$
3. 7:3, $\frac{7}{3}$
4. 24:1, $\frac{24}{1}$
5. 4
6. 8
7. 40
8. 100
9. 2
10. 12
11. 63
12. 24
13. 45
14. 6 gallons
15. $0.48
16. 40 base runs
17. $7.20

Lesson 11/pages 23-24

1. 19
2. 3,500
3. 7%
4. 187
5. 35%
6. 1,200
7. 19
8. 200
9. 300 people
10. 50 books
11. 24 salespeople
12. 18%
13. $2,400

Lesson 12/pages pages 25-26

1. $720
2. $5,355
3. $2.25
4. $202.50
5. $907.44
6. $260
7. $5,760
8. $3,937.50
9. $321.75
10. $127,500
11. $76,500
12. $150,000

Life Skill/page 27

1. $2,220
2. $420
3. $2,960
4. $1,890

Lesson 13/pages 28-29

1. increase
2. decrease
3. 20%
4. 40%
5. 22%
6. 25%
7. 23%
8. 100%
9. 37%
10. 2,300%

Life Skill/pages 30-31

1. 4:00 p.m.
2. 4:25 p.m.
3. 3 hours
4. 7:15 p.m.
5. 8:20 p.m.

Lesson 14/pages 32-33

1. 25 pieces
2. $52.70
3. 20%
4. 15 crates
5. 1,500 students

Unit 1 Posttest/pages 34-35

1. 26.332
2. 408.574
3. 19.67
4. 4.2
5. $74\frac{7}{24}$
6. $66\frac{19}{24}$
7. $3\frac{11}{12}$
8. $4\frac{2}{3}$
9. $\frac{1}{20}$, 5%
10. .8, 80%
11. $\frac{7}{50}$, .14
12. $\frac{3}{1}$, 300%
13. .625, 62.5%
14. $\frac{3}{200}$, .015
15. >
16. >

17. <
18. =
19. =
20. >
21. 440 cans
22. 98 trees
23. $14.73
24. 20%
25. 60 students

Unit 2 Pretest/pages 36-37

1. 4 c
2. 7 min
3. 33 ft
4. 96 oz
5. 1 gal
6. $2\frac{2}{3}$ yd
7. 28 pt
8. 64 in.
9. 20,000 mm
10. 1.5 kg
11. 4,000 mL
12. 1.2 m
13. .35 km
14. 9,000 mg
15. 5.12 meters
16. 3 pounds 12 ounces
17. 4 hours 15 minutes
18. 1.25 liters
19. 16 jars
20. 850 milliliters
21. BBB Home Improvement Center
22. AAA Hardware Store

Lesson 15/pages 38-39

1. W
2. L
3. L
4. V
5. W
6. L
7. L
8. V
9. W
10. L
11. L
12. V
13. L
14. V
15. W
16. W
17. W
18. V
19. F
20. J
21. G
22. I
23. A
24. D
25. C
26. E
27. H
28. B

Lesson 16/pages 40-41

1. 28 quarts
2. 144 inches
3. 3 days
4. 48 ounces
5. 5 minutes
6. 12 pints
7. 35,200 yards
8. 1,440 minutes
9. 120 inches
10. 32,000 pounds
11. 160 quarts
12. 3,600 seconds
13. 840 hours
14. 3,520 yards
15. 32 quarts
16. 200 yards
17. yes
18. no

Lesson 17/pages 42-43

1. 4 gallons 2 quarts
2. 2 pints 1 cup
3. 20 pounds 12 ounces
4. 2 tons 500 pounds
5. 6 yards 1 foot
6. 2 quarts 1 pint
7. 3 hours 15 minutes
8. 6 days 8 hours
9. 2 feet 3 inches
10. 5 minutes 30 seconds
11. 4 ounces
12. 1,000 pounds
13. 1 quart
14. 200 pounds
15. 5 yards 2 feet
16. 2 quarts 1 pint
17. 5 hours 45 minutes
18. 4 ounces

Life Skill/page 44

1. 42 hours, $619.20
2. 44 hours, $414

Lesson 18/pages 45-46

1. $\frac{3}{4}$ gallon
2. $3\frac{1}{2}$ yards
3. $2\frac{11}{12}$ feet
4. $\frac{5}{6}$ minute
5. $\frac{1}{20}$ ton
6. $12\frac{1}{4}$ miles
7. $2\frac{1}{3}$ hours
8. $\frac{3}{4}$ pound
9. $5\frac{1}{4}$ pounds
10. $\frac{1}{3}$ foot
11. $3\frac{3}{4}$ feet
12. $2\frac{1}{4}$ days
13. $\frac{1}{4}$ pound
14. $\frac{3}{8}$ gallon

Lesson 19/pages 47-48

1. 3 quarts
2. $1\frac{1}{2}$ miles
3. 7 weeks 6 days
4. 15 minutes
5. 5 pounds 14 ounces
6. 2 inches
7. 5 yards
8. 2 yards 2 feet
9. 36 feet
10. 4 gallons 2 quarts
11. 3 tons
12. 23 gallons 3 quarts
13. $18\frac{1}{2}$ miles
14. $7\frac{1}{2}$ hours
15. 20 jugs

Life Skill/page 49

1. $15\frac{3}{4}$ yards
2. $22\frac{1}{2}$ yards, $90
3. $6\frac{1}{2}$ yards, $16.25
4. 45 ounces

Life Skill/pages 50-51

1. 1400 hours
2. 1800 hours
3. 2400 hours
4. 0715 hours
5. 0200 hours
6. 1020 hours
7. 2300 hours
8. 2015 hours
9. 2145 hours
10. 1000 hours
11. 5:42 a.m.
12. 3:20 p.m.
13. 12:00 noon
14. 6:40 p.m.
15. 11:40 p.m.

16. 12:45 a.m.
17. 1:00 a.m.
18. 5:43 p.m.
19. 10:15 p.m.
20. 9:20 a.m.

Lesson 20/pages 52-53

1. $1.48
2. 2.5¢
3. 47.3¢
4. $3.90
5. 60¢
6. 11.1¢
7. Bob's Supplies
8. 36-oz bag
9. $0.25
10. Mark's Foods
11. $2.10 a quart

Lesson 21/pages 54-55

1. C
2. G
3. H
4. B
5. I
6. A
7. 102.8°
8. 99°
9. 101.2°
10. 104.6°
11. 3.8°
12. 105°

Life Skill/page 56

1. $97\frac{1}{2}$ feet
2. $92\frac{1}{2}$ feet
3. 20 feet
4. $4\frac{2}{9}$ feet
5. $2\frac{1}{2}$ inches

Lesson 22/pages 57-59

1. meter
2. milliliter
3. gram
4. millimeter
5. centimeter
6. liter
7. kilogram
8. kilometer
9. 37,000 m
10. 4.2 kg
11. 2,000 mL
12. 15 m
13. .15 m
14. 3,400 mg
15. 2,800 mm
16. .25 L
17. 2,400 grams
18. 2,000 milliliters
19. .4 kilogram
20. 9,600 meters

Life Skill/pages 60-61

1. 250 milliliters
2. 275 milliliters
3. 5 kilograms
4. 9.36 kilograms
5. 25 patients
6. 27.3 kilograms
7. 29 kilograms
8. 156 people

Lesson 23/pages 62-63

1. 8.05 kilograms
2. 2 meters
3. 11 liters
4. 2 meters
5. 15.5 kilograms
6. 89.5 kilometers
7. 799.7 liters
8. .999 meter
9. 40.36 kilograms
10. .85 kilogram
11. 4.05 liters
12. 15.15 grams
13. 10.05 liters
14. 13.2 meters
15. 1.3125 kilograms
16. .75 liter
17. .25 liter

Life Skill/page 64

1. 47 kilograms
2. 17 kilometers
3. 2.25 liters
4. 4.55 kilograms

Lesson 24/pages 65-66

1. 108,000 times
2. 3.6 kilograms
3. 10 yards
4. $2\frac{1}{4}$ gallons
5. 270 centimeters
6. 19 miles
7. 300 milliliters

Unit 2 Posttest/pages 67-68

1. $3\frac{1}{4}$ gal
2. $1\frac{3}{4}$ lb
3. 48 in.
4. 10 pt
5. $3\frac{1}{2}$ hr
6. 7,920 ft
7. 10 qt
8. 108 in.
9. .9 m
10. 3.3 L
11. 3.8 m
12. 71,200 g
13. 7 kg
14. 3,500 mL
15. 26.125 kilometers
16. 11 pounds 11 ounces
17. 30 minutes
18. 3 feet 6 inches
19. 1.88 liters
20. BBB Home Improvement Center
21. AAA Hardware Store, BBB Home Improvement Center

Unit 3 Pretest/pages 69-70

1. 2,500 people
2. 1,000 people
3. 40%
4. 6,875 guests
5. 1,719 guests
6. Shapes

	Circles	Squares	Triangles	Total
Large	1	3	2	6
Small	5	3	2	10
Total	6	6	4	16

7. 4
8. 3:2
9. $\frac{1}{16}$
10. 25%
11. $\frac{3}{5}$

Lesson 25/pages 71-73

1. Number of Accidents in One Year
2. 10 accidents
3. the number of accidents on 4 streets in one year
4. 80 accidents
5. 45 accidents
6. 3 symbols
7. 250,000 people
8. the number of people attending soccer games in different countries
9. 1,500,000 people
10. 4 symbols

Lesson 26/pages 74-75

1. 200 cups
2. 10 cups
3. $\frac{1}{5}$
4. 2 times
5. 12 pounds
6. $4,200.80

Lesson 27/pages 76-78

1. Shapes
2. circles, squares, triangles, stars
3. 3 circles, 4 squares, 2 triangles, 5 stars
4. Shapes
5. sizes
6. Cards
7. clubs, spades, hearts, diamonds, face cards, number cards
8. clubs: 2 face, 3 number
   spades: 1 face, 3 number
   hearts: 3 face, 3 number
   diamonds: 2 face, 4 number

Lesson 28/pages 79-80

1. Circles

	Color	Black	White	Total
Circles	8	3	9	20

2. Shapes

	Circle	Square	Triangle	Total
Large	2	0	3	5
Small	3	5	3	11
Total	5	5	6	16

Lesson 29/pages 81-82

1. $16
2. $80
3. $144
4. $60 total
5. $120
6. no
7. waffles
8. $\frac{1}{3}$ cup
9. 14 eggs
10. 16 cups
11. 22 cups
12. 4 cups

Life Skill/pages 83-85

1. 150 calories
2. 72 milligrams
3. 5.8 grams
4. 4%
5. 36 grams
6. 12%
7. 16.2%
8. 238 grams
9. 1 hour 20 minutes

Lesson 30/pages 86-87

1. 8:43 a.m.
2. 2:29 p.m.
3. 5:29 p.m.
4. 10:03 a.m.
5. $4\frac{1}{2}$ hours
6. 2:03 p.m.

Unit 3 Posttest/pages 88-89

1. 1,920 complaints
2. $1\frac{1}{2}$ symbols
3. 50%
4. $\frac{1}{3}$
5. 20%
6. 1,200 complaints
7. Home Service Usage Rate Chart
8. Local Calling Area, 8-15 min, 16-40 min, 40+ min
9. rate for first minute, rate for additional minutes
10. 12.6¢
11. 11.4¢
12. $2.40
13. 3.1¢
14. 6.8¢

Unit 4 Pretest/pages 90-91

1. $230 million
2. $90 million
3. $1.6 million
4. $\frac{2}{3}$
5. $19 billion
6. 80%
7. 50%
8. 30%
9. $\frac{7}{8}$
10. 1,000 feet
11. 3 times
12. 2,400 ft

Lesson 31/pages 92-95

1. 21%
2. 14%
3. agriculture, transportation
4. services
5. Finance Insurance
6. Education of Adults in the United States
7. 159 million
8. 7 categories
9. 30%
10. 4 years or less
11. high school diploma
12. 4 years or less, 5-8 years, 9-12 years, graduate or professional

Lesson 32/pages 96-97

1. 300
2. 500
3. 1,500
4. the percent of men in different countries who completed secondary education
5. Germany
6. Italy
7. 95%
8. 85%

Lesson 33/pages 98-100

1. 120 million tons
2. 30 million tons
3. 30 million tons
4. $\frac{3}{5}$
5. $33\frac{1}{3}\%$
6. 75%
7. Sample answer: 260 million tons
8. Answers will vary.
9. 1,250 tons
10. 300 tons
11. 50 tons
12. 6 hours
13. $135,000

Life Skill/page 101

1. 1,100 kilowatt-hours
2. $80
3. $50
4. January, July, August, September; Sample answer: using the heater in winter and air conditioning in summer
5. $110

Lesson 34/pages 102-106

1. Under 25; Sample answer: Many people under 25 are students who don't have full-time jobs.
2. $6,000
3. Sample answer: The difference between income and expenditures decreases.
4. income equals expenditures; Sample answer: People 65 and older are retired and on fixed incomes.

5. 45-54; Sample answer: In this age group people are making the most money in their careers.
6. $83\frac{1}{3}\%$
7. 11%
8. public administration
9. mining
10. 5,694,200 people
11. public administration, finance
12. construction, farming
13. Sample answer: 20%
14. 1987-1988
15. 29%
16. greater
17. decrease
18. Sample answer: The difference between import and domestic car sales will decrease.
19. Answers will vary.

Lesson 35/pages 107-108

1. Boiling Points of Water
2. Feet Above Sea Level
3. Degrees Fahrenheit
4. 1°F
5. 209°F
6. 212°F

Lesson 36/pages 109-111

1. 9; 67
2. 644%
3. 33; 56
4. 69.7%
5. VCRs
6. Sample answer: The difference is becoming smaller.
7. Answers will vary. Sample answer: video rental business, since VCR sales are increasing at a faster rate
8. 8.8 times
9. 25%
10. education
11. Answers will vary.
12. Sample answer: 50 million
13. Sample answer: The increased use of computers in business may increase business growth.

Life Skill/pages 112-113

1. B
2. The units are not uniformly marked.
3. B
4. Yes, the units are not uniformly marked.
5. graph C; Sales appear to be rapidly increasing.
6. graph D; Sales are slowly increasing, but not enough to give a big raise.

Lesson 37/pages 114-116

1. The stock is decreasing in value.
2. Both stocks are increasing; B&B Toys stock in decreasing while Alpha Computers stock is increasing.
3. Answers will vary.
4. Answers will vary.
5. 3 people
6. 50%
7. music videos, movies
8. 16 people
9. during a comedy program
10. 384 people

Unit 4 Posttest/pages 117-119

1. 5 million votes
2. 17 million votes
3. 76 million votes
4. $\frac{41}{90}$
5. 120 thousand magazines
6. 90 thousand magazines
7. 200%
8. 60%
9. $200 billion
10. 5 times
11. $10 billion
12. 32%
13. marketing costs
14. Sample answer: 90%

Unit 5 Pretest/page 120

1. 4 regions
2. 255 million people
3. 24%
4. 51 million people
5. 35.7 million people
6. 56.1 million people
7. 66 million people

Lesson 38/pages 121-122

1. 410,000 GED credentials
2. 5 parts
3. under 19
4. 25%
5. 17%
6. Criminal Justice System Expenditures
7. $74 billion
8. 2%
9. corrections, police protection
10. 7%

Lesson 39/pages 123-125

1. $21,300
2. $2,700
3. $1,800
4. $8,700
5. $8,100
6. $105.4 billion
7. $18.6 billion
8. 11 times
9. $6.2 billion
10. $28 billion
11. income taxes

Lesson 40/pages 126-127

1. decrease
2. 25-34
3. 25-34
4. over 55
5. 8%
6. 17.4 million people
7. 15.24 million people
8. 2.16 million people

Life Skill/pages 128-129

1. 12%
2. housing
3. 10%
4. $468
5. $288
6. $6,480
7. housing
8. Sample answer: credit card payments

Lesson 41/pages 130-131

1. 14,280,000 operators
2. 5,040,000 more operators
3. 11,515,000 male managers
4. 6 times
5. 4,620,000 women
6. 168,000 farmers

Unit 5 Posttest/page 132

1. age
2. 55-64
3. 10%
4. 54%
5. 63 million people
6. 30.24 million people
7. 20.16 million people
8. 50.4 million people

Unit 6 Pretest/pages 133-134

1.

Scores	Tally	Frequency
61-70	|||	3
71-80	𝍸	5
81-90	𝍸 |||	8
91-100	||||	4

2. 81.7
3. 82
4. 82
5. 37
6. 8
7. 3 people
8. 20%
9. 15%
10. 12 people

Lesson 42/pages 135-137

1.

Weight in Pounds	Tally	Frequency
1-100	|	1
101-200	|||	3
201-300	||||	4
301-400	|	1
401-500	𝍸 ||	7

2.

Scores	Tally	Frequency
25-28	|||	3
29-32	𝍸 ||	7
33-36	𝍸 𝍸	10
37-40	𝍸	5

3. 7
4. 13
5. 10
6. 22 people
7. 88%

Life Skill/page 138

1. 60 people
2. comfort
3. 16 people
4. 14 people
5. 15%
6. Sample answer: stress the comfort of the shoes

Lesson 43/pages 139-142

1. 6-8 hours
2. 15-17 hours
3. 100 people
4. 0-2 hours, 12-14 hours
5. 30 students
6. 62%
7. 38%
8. Answers will vary.
9. Answers will vary.
10.

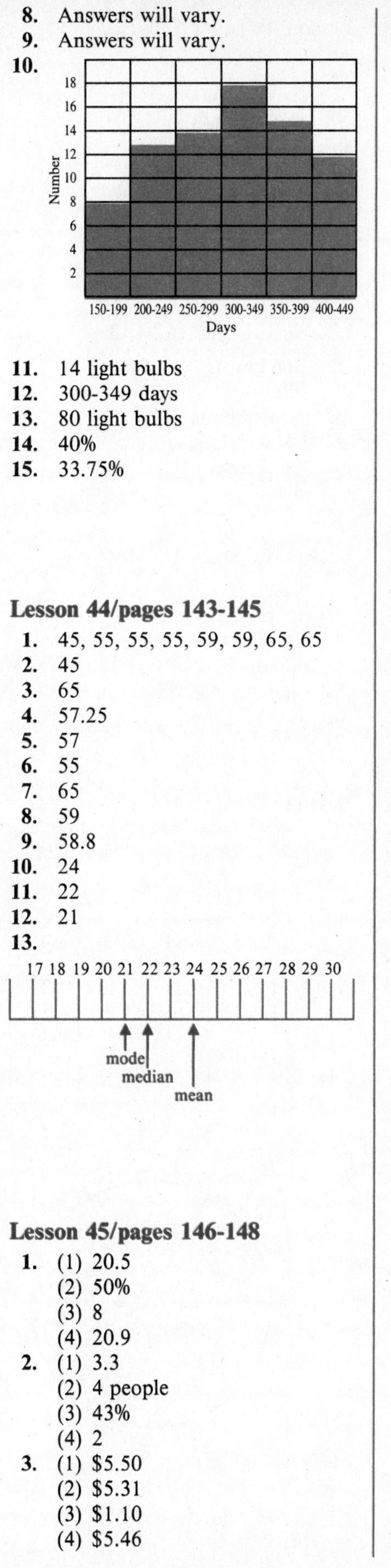

11. 14 light bulbs
12. 300-349 days
13. 80 light bulbs
14. 40%
15. 33.75%

Lesson 44/pages 143-145

1. 45, 55, 55, 55, 59, 59, 65, 65
2. 45
3. 65
4. 57.25
5. 57
6. 55
7. 65
8. 59
9. 58.8
10. 24
11. 22
12. 21
13.

17 18 19 20 21 22 23 24 25 26 27 28 29 30

mode

median

mean

Lesson 45/pages 146-148

1. (1) 20.5
   (2) 50%
   (3) 8
   (4) 20.9
2. (1) 3.3
   (2) 4 people
   (3) 43%
   (4) 2
3. (1) $5.50
   (2) $5.31
   (3) $1.10
   (4) $5.46

Unit 6 Posttest/pages 149-150

1.

Prices	Tally	Frequency
401-420	|	1
421-440	||	2
441-460	|||	3
461-480	𝍸	5
481-500	|	1

2.

Number: 1, 2, 3, 4, 5

401-420 421-440 441-460 461-480 481-500

Prices

3. $458
4. $456
5. $82
6. $461.50
7. (1) 70.5
   (2) 72
8. (1) 5
   (2) 14
   (3) 25
9. $35.17
10. $38

Unit 7 Pretest/pages 151-153

1. $\frac{1}{12}$
2. $\frac{1}{4}$
3. $\frac{7}{12}$
4. $\frac{1}{24}$
5. $\frac{1}{2}$
6. $\frac{1}{36}$
7.

A1	B1	C1	D1
A2	B2	C2	D2
A3	B3	C3	D3
A4	B4	C4	D4
A5	B5	C5	D5

8. 20 outcomes
9. $\frac{1}{20}$
10. $\frac{3}{20}$
11. $\frac{1}{10}$

12. $\frac{1}{5}$
13. $\frac{1}{5}$
14. 400 geese
15. 400 people

Lesson 46/pages 154-156

1. 1 side
2. 2 sides
3. $\frac{1}{2}$
4. 2 choices
5. $\frac{1}{2}$
6. $\frac{1}{2}$
7. 0
8. $\frac{1}{6}$
9. $\frac{1}{6}$
10. $\frac{1}{2}$
11. $\frac{1}{2}$
12. $\frac{1}{2}$
13. 0
14. 1

Lesson 47/pages 157-159

1. $\frac{1}{9}$
2. $\frac{2}{9}$
3. $\frac{4}{9}$
4. $\frac{2}{3}$
5. $\frac{4}{9}$
6. $\frac{1}{3}$
7. $\frac{2}{3}$
8. $\frac{1}{3}$
9. $\frac{2}{3}$
10. $\frac{2}{3}$
11. $\frac{1}{4}$
12. $\frac{1}{6}$
13. $\frac{1}{2}$
14. $\frac{1}{2}$
15. 0
16. 1

Lesson 48/pages 160-161

1. $\frac{1}{3}$
2. $\frac{1}{12}$
3. $\frac{1}{6}$
4. 1,500 people
5. 1,000 people
6. 1,600 people
7. low prices

Life Skill/pages 162-163

1. 1,667 birds
2. 200 deer
3. 2,000 rabbits
4. 1,000 frogs
5. 1,250 frogs
6. increasing

Lesson 49/pages 164-165

1. biased
2. no; biased
3. random
4. random
5. random; biased
6. no
7. biased

Lesson 50/pages 166-169

1. (1)

Dice	Coin	Dice	Coin
1	H	1	T
2	H	2	T
3	H	3	T
4	H	4	T
5	H	5	T
6	H	6	T

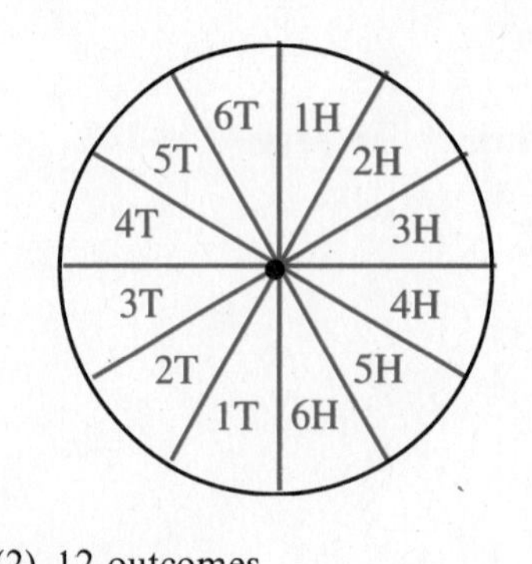

(2) 12 outcomes
(3) $\frac{1}{12}$
(4) $\frac{1}{6}$
(5) $\frac{1}{4}$
(6) $\frac{1}{4}$

2. (1) $\frac{1}{30}$
(2) $\frac{1}{15}$
(3) $\frac{1}{10}$
(4) $\frac{2}{15}$
(5) $\frac{2}{3}$

Lesson 51/pages 170-171

1. $\frac{7}{10}$
2. $\frac{1}{2}$
3. 36 outcomes
4. 75%

Unit 7 Posttest/pages 172-173

1. $\frac{1}{2}$
2. $\frac{2}{9}$
3. $\frac{25}{324}$
4. $\frac{1}{324}$
5. $\frac{7}{9}$
6.

A1	B1	C1	D1	E1
A2	B2	C2	D2	E2
A3	B3	C3	D3	E3
A4	B4	C4	D4	E4

7. $\frac{1}{20}$
8. $\frac{1}{20}$
9. $\frac{1}{10}$
10. $\frac{1}{5}$
11. 300 salmon
12. 100 people

Photo credits: Cover, © Ralph Mercer/Tony Slone Images; 8, 12, 14, 27, 43, Life Images; 44, Jules T. Allen; 49, 50, 56, 59, 60, Life Images; 64, Johnny Johnson; 82, Aaron Haupt; 85, Jeff Bates; 125, Doug Martin; 138, Glencoe file; 148, Mark Burnett; 163, Roger K. Burnard; 165, Life Images; 166, Elane Shay